A
YEAR
IN THE
BEEYARD

Other books by Roger A. Morse

THE COMPLETE GUIDE TO BEEKEEPING
BEES AND BEEKEEPING
HONEY BEE PESTS, PREDATORS, AND DISEASES (ed.)
COMB HONEY PRODUCTION
REARING QUEEN HONEY BEES
MAKING MEAD
HONEY BEE BROOD DISEASES (ed.)

A YEAR IN THE BEEYARD

Roger A. Morse

Charles Scribner's Sons   New York

Copyright © 1983 Roger A. Morse

Library of Congress Cataloging in Publication Data
Morse, Roger A.
  A year in the beeyard.

  Bibliography: p.
  Includes index.
  1. Bee culture.   I. Title.
  SF523.M75 1983   638'.1   82-42664
  ISBN 0-684-17816-1

> This book published simultaneously in the
> United States of America and in Canada—
> Copyright under the Berne Convention.
>
> All rights reserved. No part of this book
> may be reproduced in any form without the
> permission of Charles Scribner's Sons.

1 3 5 7 9 11 13 15 17 19   F/C   20 18 16 14 12 10 8 6 4 2

Printed in the United States of America.

CONTENTS

Preface	ix
1 **JANUARY:** *How to Start in Beekeeping*	1
BUYING ESTABLISHED COLONIES	2
BUYING PACKAGES	3
COLLECTING SWARMS	5
REMOVING BEES ALIVE FROM TREES AND BUILDINGS	7
BAIT HIVES	9
RACES OF BEES	11
SELECTING THE APIARY SITE	13
KEEPING BEES IN CITIES	13
THE NUMBER OF COLONIES IN AN APIARY	15
THE BEES IN JANUARY	16
2 **FEBRUARY:** *Equipment*	19
WHAT EQUIPMENT IS REALLY NEEDED?	20
FACTORY-MADE EQUIPMENT	21
HOMEMADE EQUIPMENT	24
FRAMES	27
PLASTIC BEEKEEPING EQUIPMENT	30
WOOD PRESERVATIVES AND PAINT	32

	DRIFTING	34
	DRESSING FOR THE APIARY	35
	SMOKERS AND SMOKER FUEL	37
	WHAT TO DO WHEN STUNG	38
	THE BEES IN FEBRUARY	39
3	**MARCH:** *Sources of Information*	41
	LOCAL BEEKEEPERS	42
	THE STATE COLLEGES	42
	APIARY INSPECTION	43
	ASSOCIATIONS AND THEIR NEWSLETTERS	43
	BOOKS AND JOURNALS	44
	CORRESPONDENCE COURSES	45
	SHORT COURSES	45
	CATALOGS	45
	THE BEES IN MARCH	46
4	**APRIL:** *The Start of the Active Season*	48
	UNPACKING COLONIES	49
	UNITING WEAK COLONIES	51
	THE FIRST INSPECTION—DISEASES	53
	THE FIRST INSPECTION—FOOD AND FEEDING	54
	THE FIRST INSPECTION—BROOD PATTERNS	58
	EQUALIZING COLONIES	58
	PACKAGE INSTALLATION	59
	THE SCALE HIVE	67
5	**MAY**	68
	HOW TO INSPECT A COLONY FOR CROWDEDNESS	69
	SWARM PREVENTION	72
	SWARM CONTROL	77
	CLIPPING AND MARKING QUEENS	78
	RENTING BEES FOR POLLINATION	79
	METHODS OF MOVING BEES	80
	PESTICIDES	83

6	**JUNE**	88
	DRAWING NEW FOUNDATION	89
	SUPERING	91
	DOES OLD COMB DARKEN HONEY?	92
	MAKING INCREASE	93
	WAX MOTHS	95
7	**JULY**	97
	USING QUEEN EXCLUDERS	97
	UPPER ENTRANCES	99
	SORTING COMBS	100
	VARIATIONS IN HONEYFLOWS	101
	COMB AND CUT-COMB HONEY PRODUCTION	102
8	**AUGUST**	105
	WHEN IS HONEY RIPE?	105
	METHODS OF REMOVING BEES FROM SUPERS	107
	UNCAPPING AND EXTRACTING	110
	HOT ROOMS	112
	REDUCING THE MOISTURE IN HIGH-MOISTURE HONEYS	113
	HARVESTING COMB AND CUT-COMB HONEY	114
	REQUEENING	115
	THE SECOND DISEASE INSPECTION	117
	QUEEN EXCLUDERS AGAIN— THINKING ABOUT WINTER	118
9	**SEPTEMBER**	120
	THE THIRD AND FINAL DISEASE INSPECTION	120
	HARVESTING THE FALL HONEY	121
	UNITING WEAK COLONIES	122
	WASPS AND BEES	124
10	**OCTOBER**	126
	WEIGHING COLONIES	127
	FALL FEEDING	127

	PREPARATION FOR WINTER	129
	THE BUILDING-PAPER PACK	132
11	**NOVEMBER**	135
	HONEY SHOWS	136
	PACKING HONEY	138
	LABELING HONEY	139
	RENDERING CAPPINGS AND OLD COMBS	140
12	**DECEMBER**	142
	HONEYBEE LIFE HISTORY	142
	IDENTIFICATION AND CONTROL OF PESTS, PREDATORS, AND DISEASES	145
	MICE, SKUNKS, AND BEARS	148
	HONEY PRODUCTS	151
	NATURAL HONEY	154
	BEESWAX PRODUCTS	156
	OTHER HIVE PRODUCTS	157

Selected Bibliography 160
Index 163

PREFACE

Successful management of honeybee colonies is much the same wherever bees are kept. This book is based on my more than forty years of beekeeping experience in New York State and Florida, especially in central New York and the Hudson Valley. The specific dates and timing of management practices are appropriate for New York but should, of course, be adjusted when applied to your own area.

Beekeeping is a combination of art and science. As scientists, we can study and even predict the typical yearly cycle of a honeybee colony. In central New York, the minimum population will occur around March 1 and a maximum population will be reached about July 1. Swarming (colony division) also occurs within predictable limits: Approximately 80 percent takes place between May 15 and July 15, while the remaining 20 percent occurs in a "second season" from August 15 to September 15. The availability of wild and cultivated flowering plants is dependent on the date of the last spring frost, about May 20 in my area, and the first fall frost, about September 20. A careful beekeeper fixes such dates in his mind because they dictate the management program. To be too late or too early in manipulating a colony is to jeopardize part or all of the honey crop.

The artistry in beekeeping involves being able to bring

colony populations to a peak for the start of the honeyflow. This requires recognizing year-to-year seasonal variations and responding with slight changes in management, something that is somewhat easier to accomplish in the north than in the south, where the timing of seasonal changes and colony development is less tightly intertwined and less predictable. In Florida one of the most frustrating problems is determining when the orange honeyflow will start, a date that can vary over two months. By contrast, in New York State I can predict the start of basswood honeyflow within seven days, which makes management much easier.

I have written this book to provide intelligent, practical advice on how to keep bees. For me the challenge of beekeeping is to harvest a maximum crop with a minimum investment of time and money. My apiaries are not picturesque; my combs are not uniformly free of drone comb; and not all of my equipment is well painted. I am also reconciled to the fact that the area where I live is not a primary honey-producing area. Still, I manage to harvest a reasonable amount of honey every year. More importantly, in the occasional year when conditions are perfect, I can make sure that my hives are filled with honey. At these times beekeeping is the most fun.

1
JANUARY
How to Start in Beekeeping

There are several ways to get started keeping bees. The best is to buy a secondhand colony, chiefly because there will be a crop to harvest the first year, and because it is the cheapest way to start. Another way is to buy a package of bees, usually three pounds of bees and a queen, from one of the southern states. Buying secondhand equipment in which to house the bees will save money.

Still another, and interesting, method of getting started is to collect a swarm. Every police and fire department needs a friendly beekeeper willing to pick up stray swarms, especially those that land on parking meters, fire hydrants, or in cars. Removing bees from buildings and trees can be fun, but it is too time-consuming to be profitable, though I suggest that every beekeeper do it at least once for the experience. Some beekeepers have made removing colonies from houses profitable, charging a fee to do so.

In recent years there has been a scientific study of bait hives and what bees seek when they choose a homesite. A commercial beekeeper would not find bait hives profitable in the United States, but a South African beekeeper told me he routinely collects eight hundred to one thousand swarms in bait hives each year. Cheap bait hives can be built. In a good year I have had swarms occupy 80 percent of the bait hives we had designed and positioned properly.

BUYING ESTABLISHED COLONIES

The best time to buy a colony in the north is in April or early May, although it is usually cheaper to buy colonies in the fall, but if one buys in the spring, there is no need to worry about wintering or winter losses.

The chief problem for the beginner who buys a mature colony is that he may not have sufficient information to open a colony without being severely stung. Anyone who buys bees should have some guidance and especially some aid from an experienced beekeeper. One may ask for one or more free lessons as part of the purchase price.

Prices for used equipment vary greatly. They are usually tied closely to the price of honey, going up as the price of honey increases, and also depend on the condition of the woodenware. In recent years secondhand colonies have usually sold for about half the price of a new hive and a package of bees.

When one buys bees, certain things should be guaranteed and understood. At the time of the sale the hive should be free of American foulbrood, a disease of honeybee brood. In many states a certificate of inspection is required by the State Department of Agriculture before a sale may be made or colonies moved. One may tolerate diseases other than American foulbrood, but one should know why they are present (such as a poor location) and how to cope with them.

A newly purchased colony should be queenright; that is, a queen should be present. This is best determined by looking for brood (eggs, larvae, and pupae). The new owner should determine whether the queen is young or old and in need of replacement; a large number of empty cells is the best indication that a new queen is needed. (See page 58 for more information on replacing queens.) The colony should have enough honey to sustain it, sixty to seventy pounds in the fall and ten to twenty in the spring. If the colony is to remain temporarily or permanently on land that does not belong to the beekeeper, the landowner's permission should be secured.

In recent years it has been popular to buy one-story (one

box, also known as a super) colonies brought north from one of the southern states in April or early May. These are usually good buys, though the equipment may be old and unpainted. If such colonies are properly prepared, they will have a young queen and be ready to grow vigorously.

The last question to consider when buying a secondhand colony is what to use for growth and storage supers in May and June, when the colony expands. A normal colony will occupy five to seven full-depth supers, or more half depths, at the outset of the honeyflow in June or July. Forcing a colony in one or two supers to draw foundation (new comb), because finished comb is not available, may negate much of the advantage gained by having an established colony. A colony may easily build one to ten new combs each year, but not much more.

BUYING PACKAGES

Buying package bees may be expensive, but there are several advantages, the chief one for a beginner being that the unit is small and easily managed, at least in the early stages. The more bees in a hive the more likely it is for the beekeeper to be stung. The number of bees in a package is small relative to a mature colony, and the new beekeeper gains experience as the package population grows.

In central New York State the package should be delivered and installed as soon after April 15 as possible. It is best to book orders for spring delivery in the fall, as the early bookings are filled quickly. Package bee producers are more than willing to fill orders in May and June, but this is not to one's advantage in the north. Packages installed in May and June will grow satisfactorily, but may require feeding in the fall in order to survive the winter.

The best size package to buy is one with three pounds of bees and a queen. When packages are made in the south, there is an effort on the part of most producers to use as many young bees as possible; still, the day the package is made it will contain some old bees and some will die the first day. A three-pound package contains about twelve thousand worker bees.

A three-pound package of bees as it appears when received through the mail from one of the southern states. The small can at the top of the package contains sugar syrup on which the bees feed for the two or three days they are in transit. See Chapter 4 for information on package bee installation.

The package will be in transit for one to three or more days; the queen is not likely to lay eggs the day the package is installed; and it takes twenty-one days for a worker bee egg to become an adult. Thus between the time the package is made and the first replacement bees emerge about twenty-five days later, the population drops from twelve thousand to about seven thousand. Meanwhile the quantity of brood increases and there is great pressure on the bees to maintain the brood rearing temperature of 92°F or higher. This is why I advise a three-pound package for use in the northeast.

One must not be misled by the fact that honey producers in the Peace River district in Alberta, and some other Canadian prairie provinces, start colonies with only two pounds of bees. In the northeastern states days are shorter than in Canada, and nectar and pollen flows, including those that are important for colony buildup, are less intense. So in this area three pounds of bees are needed, and even a three-pound package is unlikely to produce a surplus the first year.

Advice on the installation and management of package colonies is given in Chapter 4.

COLLECTING SWARMS

When a colony of honeybees swarms, the old queen and 30 to 70 percent of the bees leave the hive and seek a new residence. If the old queen is feeble and cannot fly, the swarm will exit with the first virgin queen to emerge from among the many the colony is growing. The swarm usually has selected a new homesite before it emerges. It settles within about one hundred yards of the parent colony, organizes itself, checks the new home again, and then moves to it. If a new home cannot be found, the swarm may move to another temporary settling site. Often a swarm is forced to settle temporarily before reaching the new home because the queen is exhausted. At any of the settling sites it is a simple task to capture the swarm.

A word of caution is in order here. When a swarm emerges, most of the bees are fully engorged. Bees full of food are much like well-fed people—they have good dispositions—and usually they will sting only if squeezed. However, swarms that have been away from their parent hive for some time may have exhausted their food. Such swarms are called dry swarms, and the bees in them may be vicious. The severe stingings one hears about occasionally result from an encounter with a dry swarm.

Attention to the weather will usually make it possible to predict if a swarm is out of food. When a swarm is caught clustered away from the parent colony in bad weather, it will be unable to forage. After a period of inclement weather, bees in a dry swarm will replenish their food, but this may take a day or more. Thus caution should be exercised in attempting to capture a swarm after the weather has been such that bees could not fly on one or more days.

A swarm may be captured by holding an empty hive (with a bottomboard in place) under the swarm and vigorously shaking the branch it is on, forcing the swarm to fall into the hive. It is best if the bottomboard is stapled or nailed into place ahead of time. The hive should then be quickly placed onto the ground, the frames (combs) put in place, also quickly, and the cover put onto the hive. If the queen is in the box, the swarm will usually accept the new home. Bees do not care for a nest exposed to the sunlight and may abscond (exit) if the hive is in the sun. The bees and their new home should be placed in a shaded spot within about twenty-five feet of the place where they were clustered. After dark the entrance should be closed, and the colony moved to its new location.

An alternative that will work (if one is very careful) is to cut the branch on which the swarm is clustered, carry it to a new homesite, and dump it into the hive in the same manner. I have carried swarms several hundred yards in this way. If some of the bees fall off, place the branch just over those on the ground and let them recluster. This takes time and patience.

Still another method is to shake the swarm into a sack or bag. Plastic should not be used, as the bees may suffocate. With care, and luck, one shake will dislodge the bees and almost all will be in the bag, to be carried, for an hour or even a day, wherever one wishes.

It is more difficult to capture swarms on an object such as a parking meter, a phone booth, or some similar object. If one places a hive entrance immediately adjacent to the swarm, it may move in; a frame of brood is a strong attractant. More often it is necessary to work one's fingers through the swarm in hopes of finding the queen. If she can be found and caged, the bees will not abandon her but will cluster around her, or move into a hive if she is placed inside.

Hiving a swarm is a cheap way to obtain bees, but the problem is the same as hiving a late package: The bees may need to be fed to survive the winter.

REMOVING BEES ALIVE FROM TREES AND BUILDINGS

Removing colonies alive from trees and buildings is an unprofitable but worthwhile experience that every beekeeper should have at least once; it certainly gives one a better appreciation of the movable-frame hive. The chief difficulty in removing the nest and keeping the bees alive is the way in which the combs are intertwined and attached to the sides of whatever surrounds the nest. A few beekeepers who are also carpenters have a profitable sideline of removing nests from houses and then repairing the inevitable damage to the structure.

It is important to emphasize that if one is to eliminate a nest of honeybees from a house or building, the comb, honey, and refuse that accumulate under a nest must also be removed. If not removed, these materials will attract a host of insects and rodents that may prove to be a worse problem than the bees. Most importantly, the area occupied by the nest must be packed tightly with insulation, sawdust, or some other material so that the bees cannot occupy it again. The odor in such a location will cause bees to be attracted there for years, even decades.

A simple but time-consuming way of removing bees from a natural nest is to plug all the entrances but one and to put over it a cone of wire mesh with an opening at the tip of the cone just big enough to allow one bee at a time to escape. A nucleus colony (a small colony containing a thousand or more bees, a queen, and some brood) is placed within a foot or two of the cone. As bees forage from the natural nest and are prevented from reentering their home, they will join the small unit. After several weeks the bulk of the bees will have been captured and the remainder of the original nest can be removed. Usually the queen in the original nest is lost.

One removes bees from a tree in a slightly different manner. The tree is cut down and the section containing the active nest cut out of the trunk. Determining the upper and lower limits of the nest is not always easy, though one may be guided by watching the saw blade for signs of wax and/or honey. In a nest that has been established in a tree for more than a year the combs will be sufficiently reinforced and tough that they will rarely break when the tree falls to the ground. The portion of the tree with the nest is then placed into an upright position and the portion above the nest removed so the top of the nest is fully exposed. This may require cutting away some of the comb. A piece of plywood about sixteen by twenty inches, the approximate outside dimensions of a standard ten-frame super, is cut and a hole the size of the exposed portion of the nest is made in the center of it. The hole and the exposed portion of the nest should be at least six inches in diameter for this method to work. An empty super of combs with a cover is then placed on the plywood above the nest in the hope that the bees and the queen will move upward, which may take a few days or several weeks. When brood is found in the super, a queen excluder (a screen that allows the passage of workers but not the queen bee) is placed under it and on top of the plywood. If one is lucky, the queen will be caught above and the brood below will emerge, freeing the bees of their original home. One can then remove the super, which now becomes the primary brood nest, and put it in a permanent location at one's convenience.

When removing bees from a house or building with an interior finish, the nest may be entered from the interior or exterior of the building. But if one is to find the queen and take the bees alive, smoke will be needed, and rarely can smoke be used inside a house without complications. When one is merely interested in removing bees that are a nuisance and the bees are first to be killed with an insecticide, it is often easier to enter the nest from the inside of the building through the plaster or plasterboard. Often it is cheaper to repair the damage to an inside wall than to rebuild the outside of a house.

Removing the outer wall of a building and removing combs one by one in search of a queen is a challenge. When the queen (which is much larger than the other bees) is found, she is caged and placed in a hive body, with frames, adjacent to the nest. Although combs in the natural nest may be saved by tying them into an empty frame, and while doing so may appear to be prudent, it is not worth the effort. Good, uniform, strong combs are wanted in frames in hives, and these can be obtained only by using wired frames with foundation. The comb, as it is removed, is put into a plastic bag or other type sack where the bees cannot find it. When the bees are denied their own comb and brood, they will join their queen. If one places a frame with a small amount of brood in the hive, either from another hive or the nest, the bees will move in more rapidly. In the evening all the bees will enter the hive, and it may be closed and carried where it is wanted. One should then wash and clean the old nest site and pack it with insulation so that bees cannot use it again. I prefer to let a professional carpenter repair the building.

BAIT HIVES

A bait hive is a box hung in an appropriate place for the purpose of attracting and capturing a swarm. When bait hives are properly built and placed, in central New York one can expect 50 to 80 percent occupancy in the course of a season. The most active swarming season is May 15 to July 15. Since bees scout new homesites before the swarm leaves the hive, it is important to have the bait hives in place a week before May 15.

Using bait hives is not a new idea. Native East African beekeepers have been using them successfully for several thousands of years. A recent scientific study of bait hives provides information about the design, type, and location of the bait hives that bees prefer.*

Bait hives can be a cheap source of bees. Swarms captured earlier in the season will be more valuable than those captured later, just as early packages are more valuable. Late swarms will probably need fall feeding to survive the winter. Swarms should be removed from the bait hives as soon as possible. I have not had much success overwintering bees in bait hives, both because they have often not had time to store sufficient food and because a good location for capturing a swarm may

---

*Bait Hives for Honey Bees, Information Bulletin 187 (8 pages, 1982), by T. D. Seeley and R. A. Morse, can be obtained for a small fee by writing Mailing Room, Building #7, Research Park, Cornell University, Ithaca, N.Y. 14853.

Paired bait hives in a lure test. The bait hive on the right attracted a swarm. These hives, facing west, were well shaded.

not be a good wintering site, since a bait hive is not as sturdy as a tree.

The studies Seeley and I made show that bees prefer a box about the size of a ten-frame Langstroth super (see page 24); in fact, an old super with pieces of plywood nailed on the top and on the bottom makes a good bait hive. An eight-frame hive will also serve. Bees prefer an entrance about $1 1/4$ inches in diameter; it should be near the bottom of the nest for best results, though bees will sometimes accept a nest with an entrance near the top. We have never used wood less than five-eighths of an inch thick and suspect, though we have no data, that bees can measure and will reject something less solid. The top must be secure and no light must enter around it. One may tape wood joints if the joints are not square and light tight. One must not put combs or frames into a bait hive, as they may be attacked and/or destroyed by wax moths, rodents, and a host of insect pests.

To be successful, bait hives must be elevated. One year we had forty bait hives of good design on stands thirty-six inches high and only three were occupied. Bees seem to prefer to nest about ten to fifteen feet above the ground. The nest box should be visible but shaded. We have had bait hives accepted by swarms during cloudy weather or at a time of day when the hives were shaded, only to watch them be vacated when they were hit by the sun. At one time we had bait hives on dead trees, but these provide no shade and we now find it best to use only trees with large spreading limbs.

One serious problem is taking a full bait hive down from a height of ten or fifteen feet. During a honeyflow a bait hive can gain weight rapidly; we have had bait hives that weighed thirty to fifty pounds when filled, which can be cumbersome objects to manage on a ladder. We usually fix our bait hives in place with one or two nails. Before removing the nails, we tie a rope around the bait hive and put it over a limb so that when the nails are removed, the bait hive can be lowered to the ground easily.

RACES OF BEES

Honeybees are native to Europe and Africa and were carried to all continents by the early European settlers. In Europe there are almost as many races of honeybees as there are languages of people.

I believe that the best race for our purposes in North America is the Italian bee. These bees show good resistance to several common bee diseases, especially European foulbrood, a bacterial disease. They are reasonably gentle, use little propolis,* form a compact brood nest, and winter easily. In my opinion Italian bees are the best honey producers.

Caucasian and Carniolan bees, from eastern Europe, are

---

*Definitions of *propolis* and other terms and discussions of the life history of the honeybee and bee diseases can be found in Chapter 12.

gentler, but are not such good producers and use too much propolis. In fact, I am annoyed with the amount of propolis in hives today, which is a direct result of efforts to grow and sell gentle bees. When I was a boy, I did not get my fingers stuck up with propolis as I do now after working only a few hives of bees. The excessive propolis in hives today is not necessary and it slows up hive manipulations. It is difficult in some hives to search for a queen and, having found her, to pick her up to clip her wings, because one's fingers are so sticky.

A fourth race that one sometimes hears about in America is the German black bee. I know little about it and see it only rarely. These bees have the peculiar habit of running on the comb and even falling off onto the ground. They are good honey producers and winter well, but have gone out of favor because of their susceptibility to European foulbrood.

However, more important than the race of bees one selects is the beekeeper who grows the queens. Queens vary greatly, as do most animals. The most important time in their life is their five-and-one-half-day larval, or growing, period. The best queens are those that were well fed, in a properly controlled environment, as larvae. Growing queens when there is an abundance of pollen to provide the protein they need is especially important. It is equally important to have good conditions for mating. Queens and drones mate twenty to seventy-five feet above the ground, sometimes a mile or more from their hive. Obviously, the weather must be good, and in the early spring this may not be the case. A responsible queen producer will sell only queens that he knows are properly mated. Poor mating is probably a chief cause of bees superseding (replacing) their queen.

Many claims are made for this or that type of queen, but as far as I am aware there are no data to support the notion that any one selection on the market today is superior. This situation could change rapidly; in fact, I am impressed with some of the information I have seen on certain disease-resistant queens. All longtime beekeepers have a favorite queen breeder. I have noted that many, though obviously not all, of the best queen breeders have the small ads in the bee journals, since they are well known among beekeepers and do not need to advertise widely.

SELECTING THE APIARY SITE

If a swarm of honeybees is given a choice of homesites, it will prefer one that faces south or east. I know, however, that the honeybee is an adaptable animal and can be kept under a variety of conditions; in this regard it is important that the apiary site suit the beekeeper too.

A good apiary site is well drained, protected by hills or trees from harsh winds, exposed to the sun (contrary to choices made by swarms), level so that working in the location is easy, slopes slightly to the south or east, has a good access road or path, has water nearby, and, especially important, is in a place where the bees will not be a nuisance.

KEEPING BEES IN CITIES

Bees are kept in almost all of the major cities in the world, usually without any problems. One must always be sensitive to a neighbor's concerns, but if the bees are properly managed there will be no problem. In cities and villages there are almost always good foraging areas for a certain number of nectar-feeding insects. The common plants include clovers and dandelions in lawns, flowering shrubs and trees, ornamental flowers and weeds in gardens, roadside plantings, and sometimes even rooftop gardens. I am often asked to testify or write a letter giving an opinion about banning the keeping of bees in a city or village. I point out that if people are really serious about ridding an area of stinging insects, they must first get rid of their food. It is a simple matter to pass a law prohibiting beekeeping in an area, but if the man-kept colonies are removed, they will be replaced by wild colonies living in trees and buildings and by wasps, especially yellow jackets, all of which will take advantage of the food supply.

I advise that when bees are kept in a congested area, the hives should be surrounded by a hedge or fence so the bees are forced to fly up in the air and above the heads of people in the vicinity. If pedestrians are forced to walk through a colony's flight path, they might be stung, not because foraging bees are aggressive, but merely because the bees may accidentally hit a person, making it a nasty encounter for both.

The city beekeeper should be especially aware that some races of bees are much gentler than others. Caucasian bees are especially suitable for city beekeeping, though with some searching one can find Italian bees that are just as gentle. In a city one must be more careful when manipulating colonies. A commercial beekeeper who has much work to be done will often work his bees on a day when they are not foraging and as a result there will be more and more angry bees in the air as he proceeds. A city beekeeper should therefore work his bees only on warm, sunny days, move slowly, and always be careful to use the right amount of smoke.

Bees use water to cool the hive on hot days and also to dilute honey fed to larvae; they will usually take water from the closest source. In a city it is always advisable to provide bees with water so they will not gather it from a neighbor's birdbath or swimming pool. The water need not be fresh, but there should be a float on which the bees can land. A can filled once a week is usually adequate.

One especially difficult problem for the city beekeeper is that his bees must defecate somewhere. Honeybees do not void their feces in the hive under normal circumstances. If they do, social order breaks down rapidly and the colony is doomed, usually within a few days. When bees have been confined for a week or more, they may accumulate a great deal of fecal matter, though much depends on the type of food they have been eating. Bees fed sugar syrup or light honey will have much less fecal matter than will bees feeding on dark honey. In a city, bees filled with fecal matter may fly and void feces over an area a quarter mile or more in diameter. Little brown spots may appear on cars, lawn chairs, windows, and, worse still, on clothing put out to air or washing hung on a clothesline to dry. In fact, the home clothes-drying machine has done much to eliminate a serious problem for city and village beekeepers. Fortunately, these brown spots disappear from other objects during the first rain.

THE NUMBER OF COLONIES IN AN APIARY

Beekeepers do not grow plants for honey production, because it is just not possible to provide the bees with all the forage they need. One might plant a field of buckwheat and hope the weather is such that it might gather a crop of honey; however, planting a crop does not guarantee it will produce nectar. Such a venture is profitable only if there is someone with the necessary machinery to harvest the grain and, after that, someone to buy it. A successful beekeeper moves his colonies to those areas where honey plants abound or he comes to understand the limits of his territory and keeps only as many colonies as the area will support. Almost any site in the northeast will support one colony of bees, though there are areas in the mountains where this may not be true. I do think it is helpful to have some honey- or pollen-producing plants around one's home and/or apiary, as it keeps one in touch with what is taking place. Commercial beekeepers feel an apiary should be capable of supporting at least thirty colonies to be practical.

From time to time maps have been prepared outlining the better and poorer beekeeping areas in a state. Many times these have been published in the bee journals available in some libraries. In New York State, for example, the general information bulletin on beekeeping has a map showing the primary, secondary, marginal, and submarginal beekeeping areas in the state. Some of the state maps made twenty or even fifty years ago are still valid, though one must be aware of changing agricultural practices. In several northeastern states, for example, many areas that once produced primarily buckwheat honey have gone out of buckwheat production. Some of the fields where buckwheat once grew are now covered with goldenrod, which may yield nectar under some circumstances; other fields in the area are now woods; and some areas contain basswood trees that may produce a crop in some years. The professor of beekeeping at the state college, the chief apiary inspector, or the state entomologist should be familiar with what has been researched and published in the past.

THE BEES IN JANUARY

Two excellent studies, one by Dr. Edward Jeffree of Scotland in the 1950s and the second by Professor Alphonse Avitabile of Connecticut in the 1970s, show that normal colonies in a northern climate rear the least amount of brood in October and November. Most colonies will have no brood during those two months. In addition to these studies, I have examined a small number of colonies in December. Eggs can be found during most of the month, but only a small percentage of colonies will have larvae; I do not know whether the eggs fail to hatch or whether they or the young larvae are eaten by the worker bees.

In any event, it is in the month of January that nearly all colonies will be rearing brood. It is important that the colony interiors be dry. Honeybees have no difficulty maintaining a brood rearing temperature of 92°F or above if they are dry, or at least if the brood rearing area and above are dry. When bees generate heat, they consume a great amount of honey and in the process give off a great deal of water. They must therefore be able to ventilate their hive to rid their colony of this water. Proper methods of preparing colonies for winter are discussed in Chapters 10 and 11.

In January a chief concern is that colony entrances remain open to permit flight on those rare days when the bees might fly, even if for only half an hour. Honeybees do not void fecal matter in the hive under normal circumstances, and thus these short winter flights are important. The greatest danger in winter is that entrances may become plugged with dead bees or ice.

Many beekeepers like to provide their colonies with an upper entrance. This is most often done by drilling a $1/2$-inch hole above the handhold in the top super. I favor the idea, though I know of no data to support the theory that colonies with upper entrances winter better.

Dead bees on the snow in an apiary in winter. This is a common sight and a sign that the bees have had a good flight. The bees that are lost are the old ones. There are many specks of fecal matter on the snow and hive cover on the right, again a good sign.

The debris under natural nests in trees may be wet throughout the year without affecting colony survival, but nests in trees are often far above ground, which may aid the bees in winter. I suggest it is important to do all one can to keep bottomboards dry in the winter. One should not disturb a colony in the winter, other than to clean an entrance; however, in January one can make observations and notes of items to be corrected another season.

If one uses bottomboards with an alighting board that projects in front of the hive, ice formed by water dripping off the hive cover or packing may accumulate there. The hive should tip forward sufficiently so that water that drops from the cover or packing falls in front of the hive. An alighting board in front of a colony is not necessary. Factory-made bottomboards are usually twenty-two inches long, two inches longer than the standard super. Many commercial beekeepers use bottomboards that do not project beyond the hive at all.

Dead Bees on the Snow

One will often see many dead bees, sometimes hundreds, in addition to brown spots of fecal matter, on the snow in front of colonies. This is a normal situation in winter, especially in January and February. If these bees are examined closely, it will be noted that usually their wings are frayed and they have lost much of their body hair, which are both signs of old age. I do not know if old bees purposely take flight to die outside or if they fly to void fecal matter and are too feeble to return. In any event I feel their loss is normal and that it is better that they die away from the hive than inside, where an accumulation of bees could plug an entrance or cause a disagreeable odor.

Colonies in Snowbanks

I have never had a colony die as a result of being buried in a snowbank, nor have I dug bees out of a snowbank, though I have known of beekeepers who have done so with no apparent damage to the bees. A snowbank probably offers great protection against the wind, acts as insulation, and yet is sufficiently loose that bees can ventilate the hive.*

I presume that farther north than New York State colonies might suffer if they were confined in a snowbank too long. Still, so far as I am aware, no one in Canada advocates routinely checking and digging colonies from snowbanks. Beekeepers who overwinter bees in the Canadian prairie may face special problems in this regard, but they also feed their bees sugar syrup exclusively for the winter, for bees fed on sugar syrup accumulate less fecal matter than those fed on honey.

*Bees in colonies must ventilate to rid the hive of carbon dioxide as well as moisture. Considerable quantities of carbon dioxide may accumulate as a result of normal metabolism within the hive.

2
FEBRUARY
Equipment

I am disturbed and sometimes angered when someone sends me, or I see in the bee journals, plans for a new type of beehive. I do not know how many beehives have been invented in the past hundred or so years, but they probably number well over a thousand. It has all been a great waste of time, talent, and money.

The modern beehive, which almost all beekeepers in the United States use, was invented by Reverend L. L. Langstroth in 1851. What Langstroth discovered was that if one leaves a passageway or walking space, which we call bee space, around and between the combs, then one has a movable-frame hive. If the bee space is too wide or too narrow, the bees will close it with comb or bee glue (propolis) and one will not be able to manipulate the combs, which become stuck into place like combs in a bee tree.

The Langstroth super (hive) is the depth it is because that was the width of the board Langstroth had when he built his first hive. The hive is the width and length it is because boxes of that size were convenient for Langstroth to carry. The fact that honeybees have survived so long in hollow trees of varying sizes and shapes, as well as in the great number of beehives designed by man, indicates they are an adaptable animal: They can be kept successfully under a variety of circumstances. For

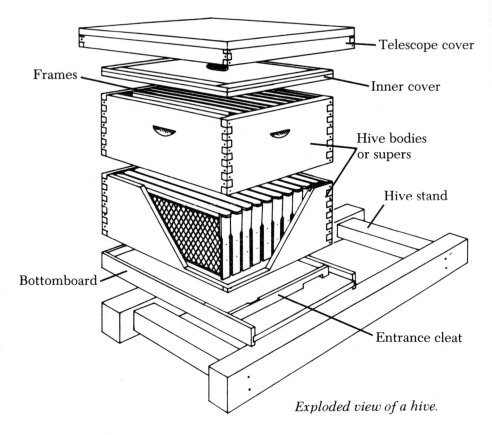

Exploded view of a hive.

this reason most beekeepers choose a hive that is simple to make and use.

WHAT EQUIPMENT IS REALLY NEEDED?

To manage bees successfully one needs only five or seven pieces of equipment: a bottomboard, a hive body (super), frames to fit the super, a cover, and a queen excluder, making a total of five. Those who prefer shallow supers because they are lighter to carry will want two more pieces, the shallow supers and frames to fit them, totaling seven. In each case, the simplest design is the best. In addition, some kind of hive stand, such as cinder blocks, railroad ties, bricks, or whatever is available at a low price, is advisable. Many beekeepers like to use inner covers, and I admit they are useful with the telescope-type covers I like to use in the north; however, if I were to move bees frequently, I would soon get rid of the telescope covers and inner covers and settle for something less bulky.

One point is paramount as far as equipment is concerned: All the parts should be of a standard dimension so that they can be interchanged and their resale value maintained. Most of the equipment in use today is of the standard ten-frame Langstroth dimensions. Perhaps 5 percent of beekeepers use supers that hold eight frames and I will admit these have some advantages. Any other size or design of equipment should be avoided.

Two other items that may prove useful during the year are division board feeders (see feeding, page 54) and entrance cleats or entrance reducers (see drawing of exploded hive).

FACTORY-MADE EQUIPMENT

Most of the manufacturers of beekeeping equipment in the United States make a good product that will give many years of service with proper care. However, one gets what one pays for and, generally speaking, the lower the price the lower the quality both in the accuracy of the measurements and saw cuts and the type of lumber used.

Properly nailing woodenware will do much to increase its value and longevity. Bee supply manufacturers sometimes provide too few nails and always nails that are too short. The longer a nail, the greater its holding power. As a general rule one should buy nails about one-quarter longer than those sold with knocked-down equipment. I prefer box nails for nailing beekeeping equipment; they have a smaller diameter than common nails. The problem with box nails is that they bend over easily when driven in place, which is a nuisance; however, one is much less likely to split wood using box nails and this makes up for their inconvenience. In nailing frames it is especially important to drive the nails in far enough so that the heads are slightly below the surface of the wood; in this way they will not catch the knife blade when frames of honey are

This factory-made cover appears to have been well nailed, but has pulled apart as the unpainted wood weathered. If repaired promptly, it should still have many years of use.

being uncapped. I do not use glue when nailing up equipment, but I am aware that many commercial beekeepers feel it is worth the effort.

The dimensions used by different manufacturers are only slightly different from each other. Most wooden parts for a single piece of equipment, such as a frame, are not interchangeable; however, frames from different manufacturers are sufficiently alike that frames from different firms may be used in the same super.

Most factory-made hive bodies have predrilled nail holes. These are usually accurately and carefully done and often prevent wood from splitting. I find that predrilling certain other hive parts, such as the side rails of bottomboards, is also worthwhile. One should also check the quality of the wood used on other hive parts, even frames, before starting to nail these together and sometimes predrill them. The drilled hole should be slightly smaller than the diameter of the nails being

Before and after repairing and painting of factory- and homemade supers. It will be noted from the unpainted supers that the weakest points are the top corners, which will pull apart if not well nailed. Note also that the corners of the homemade supers are butted together with a simple but satisfactory joint.

used. Pine is used almost exclusively for making beehives and hive parts, but there are many kinds of pine, and, even within one type, many grades.

HOMEMADE EQUIPMENT

Most commercial beekeepers make their own equipment, including the frames. Doing so allows them to use their own time, and any spare time their help may have, constructively. There is no reason why beginners or hobby beekeepers should not do so too and perhaps save themselves considerable investment.

If a beekeeper is to make his own equipment, he should understand how bee space works. It is also essential to understand the importance of standardization and to be careful to have the proper dimensions.* I suggest that one who makes his own equipment should have the standard dimensions written and nailed to the shop wall. Too often I have known of beekeepers who have pulled a super off a stack to use as a standard, only to pick a super whose dimensions were off standard.

Commercial manufacturers of equipment use metal rabbets for frame rests in their supers. They do this in the belief that the bees will be less inclined to deposit propolis on the metal, thus making it easier to manipulate the frames. My experience is that this is not true and that having a metal rabbet makes little or no difference. Beekeepers who make their own equipment do not bother with metal rabbets, but it is important to cut the rabbet to the proper depth.

*Plans and Dimensions for a 10-Frame Bee Hive (one page) is available at no charge from the Bioenvironmental Bee Laboratory, USDA, Plant Industry Station, Beltsville, Md. 20705. The most important dimensions are the outside measurements for frames, $17^5/_8$ by $9^1/_8$ inches, exclusive of the lugs, and the inside measurements for supers, which are $18^1/_4$ by $14^5/_8$ by $9^5/_8$ inches. I use the inside measurements for supers, since lumber varies in thickness, though I think of the outside dimensions as being 20 by $16^1/_4$ inches if the lumber is three-quarters of an inch thick. No two manufacturers make equipment that is precisely the same. One will sometimes find slight variations in dimensions given in different textbooks. Since there is a slight variation in bee space, one can tolerate some minor differences.

Simple telescope covers. These may be made using scrap lumber and/or plywood that may be covered with metal or roofing paper. Note the way in which the metal is bent down over the corner so as to shed water.

Homemade, lightweight combination cover–bottomboards (in one position a cover and in the other a bottomboard). These were made using wood shingles with a piece of heavy roofing paper in between and were designed by Florida–New York migratory beekeepers who wanted to reduce the weight of the load they transported. They are about one-eighth of an inch shorter and narrower than a standard hive body so that they will not project over the edge of the hives and be in the way. These cover–bottomboards are flexible and will withstand rough use.

Still another consideration in making one's own supers is the handhold. One should not skimp on these, for a super full of honey is difficult to hold without handholds of the proper depth, width, and length.

In recent years I have seen plans advocating the use of plywood to make supers and hive parts. Plywood may be used satisfactorily for covers to be covered with galvanized iron, aluminum, or tar paper, but not for much else. Some commercial beekeepers who make their own pallets may use plywood for the bottomboards, but pallets get rough usage and do not have a life of more than ten years, so the plywood makes little difference in this regard.

In no event should plywood be used to make supers. Even the best plywood will pull apart within ten years. There are many solid pine supers in the Cornell apiaries that are fifty or more years old. This is not unusual, and there is no reason one should not expect this kind of life from properly made beekeeping equipment.

FRAMES

In this section I will discuss two points: the proper nailing and wiring of a frame and the manufacture of homemade frames. Frames take a great deal of abuse and must be well made in order to have a long life. Once a frame has been in use a few years, there is a sufficient accumulation of cocoons, and propolis has been added in a large enough quantity that the comb can withstand rough treatment. However, until that time the frame's strength and the quality of the comb it surrounds depend on the wood out of which the frame was made and the care with which it was assembled.

Frames should be made using ten wire nails. The manufacturers recommend 1- or $1^{1}/_{4}$-inch nails for the top bar, but I use these to nail the bottom bars. I discard the shorter nails they advise for the bottom bars, and buy threepenny nails or $1^{1}/_{2}$-inch-long wire nails to nail the top bar in place. There is no question that my using the longer nails results in my sometimes splitting the wood on some frames. This is a loss, but if I

Frames. Wire is used to hold the foundation in place in new frames. The wiring machine shown has a handle on the right that is used to pull the wire tight. One uses two wires when the foundation is pre-embedded with vertical wires; four wires are used with unwired foundation.

am careful I do not do it often and the frames I then have are stronger. Some of my friends glue their frames in addition to using nails. I have no objection to this practice, but I do not think it is necessary if one nails a frame properly.

Whether one uses wired or unwired foundation depends on how much time and money one has. The cheapest way to make a good frame is to use horizontal wires (four for a full-depth frame) and unwired foundation. I like to use metal eyelets in the end bars so I can draw the wire extra tight. The foundation is probably held more rigidly in place by using two horizontal wires (eliminating the top and bottom wires) and using foundation with vertical wires already embedded; however, I do not think the difference is really too great. Whichever method is used, I believe it is best to heat and embed the horizontal wires into the foundation. I do not care for spur embedders (those that merely push the wire into the wax) or weaving the foundation between the wires. Both methods are used, but I have noted that most commercial beekeepers prefer an electric embedder. In any event, the foundation should not be put into the frames until just before the foundation is put onto the hive. This is discussed in more detail in Chapter 5.

Beekeepers with high labor costs, or little time, prefer foundation that is made with both horizontal and vertical wires embedded by the foundation manufacturer. I have no objection to this type of foundation—it is just that it is much more expensive.

Foundation comes in various thicknesses and weights. The heaviest available should be used for brood combs despite the fact that it costs more. The advantages and disadvantages of plastic foundation are discussed later.

One can save much money by making one's own frames. It is important to select a soft wood that is easily nailed. The native white pine grown in the northeastern United States is, in my opinion, the best pine in the world for frames and hive parts. It takes a nail easier than any of the soft pines from the northwestern states, though lumber from this part of the country is next best. Rarely will one find frames made from basswood or some other soft wood. Hardwoods, of course, while they make stronger frames, cannot be nailed easily.

A favorite frame is the free-hanging frame. This is a frame that has a top bar, a bottom bar, and end bars of the same width. I have used various widths, from three-quarters of an inch to one inch. I like a free-hanging frame that is seven-eighths of an inch wide. A narrower frame is not as easy to hold when uncapping and a wide frame is too clumsy. Free-hanging frames are a nuisance because they must be spaced by hand, but they make it possible for the bees to ventilate the hive with ease. One problem with the factory-made frames is that the top bar is too wide, reducing ventilation. Wide top bars were invented in the comb honey era, about a hundred years ago. They act in part like an excluder and tend to slow down a queen's movement from one super to another. Most beekeepers I know use only nine frames in the brood nest of a ten-frame hive, chiefly because it is easier to manipulate the frames; however, when one has frames with wide top bars, using only nine frames also improves the ventilation.

Many commercial beekeepers in the west who prefer half- or three-quarter-depth supers for extracted honey production use narrow top and bottom bars, but wide end bars that are self-spacing. They often make the frame so there is no bee space between the end bar and the super. This is a good way to make a strong frame and it does save time when one is not forced to space the frames. There is great variation possible in building a frame; the most important point is to make it sturdy.

PLASTIC BEEKEEPING EQUIPMENT

I have never been enthusiastic about plastic, either for beekeeping equipment or elsewhere; however, in almost anything we think or do there are exceptions and this is true here also. Equipment for round comb honey sections and well-made plastic foundation are two items that are acceptable.

The chief problem with plastic is that it has a short life. Very few plastic items are usable after five or ten years. If one is careful, there is no reason why wooden supers, inner covers, covers, and even some frames should not last fifty years or more.

There are some metal-covered, wooden telescope hive covers still in use in the Cornell apiaries that were made in 1902. We know the year because W. L. Coggshall, the New York State beekeeper who made them and who owned over three thousand colonies, put several sheets of newspaper between the wood and the outer metal cover. His son donated many hives and these covers to Cornell University when he retired in the 1960s. When we were repairing some of these covers a few years ago, we found the newspapers, which made for enjoyable reading. If one has equipment with that kind of life, one's investment is secure.

Round comb honey sections can only be made with plastic. It is just not practical to make wooden furniture for a comb honey super that will hold round sections. The idea of making a round section was originated by Dr. Wladyslaw Zbikowski, who died in 1969; he called his sections "cobana" comb honey sections. After he died, two firms marketed supers to accommodate these sections and they have since become very popular. Their advantage over the traditional square or rectangular sections is that they eliminate the problem of bees not filling in the corners. In fact, bees do such a good job in filling and capping round sections that they are unfair competition for the standard sections in a honey show. Nothing has a flavor superior to a fresh section of comb honey and I recommend that anyone wanting to produce a unique natural product consider its production. I have no data on how long a life a round section super might have, but I have seen some in service that are a little over five years old.

A second piece of worthwhile plastic equipment is the foundation made by Arnaba Ltd. of Hawaii. Many people have tried to reinforce honeycomb in many different ways, the most popular method being to embed wires in the beeswax foundation. A metal comb was invented and widely sold in the early 1920s; it did not remain popular for long, one reason being that the cells were easily dented. Wood splints, strips of metal, and even a foundation with a midrib of aluminum have been sold at one time or another. Several companies have made plastic foundation, but there was always a problem: The cells' base walls were not sufficiently sharp to enable the bees to draw the cells satisfactorily. Some types of plastic foundation were coated with beeswax; these worked quite well when there was a strong honeyflow. However, in a weak flow the sheets of foundation were only partially drawn; in undrawn regions of the comb, the beeswax coat on the plastic core would be chewed off by the bees during a dearth and then the bees, with no cell edges to guide them, would build extra unwanted comb (called burr and brace comb), often perpendicular to the plastic foundation. The only plastic foundation that commercial beekeepers have told me is satisfactory is that made by Arnaba. This foundation is being used more widely each year where labor costs are high. Arnaba has been widely used for only a few years and at this writing I have no data as to the length of life it might have. It is possible that since it is covered with wax, the combs may last more than the ten years I would expect for most plastic.

WOOD PRESERVATIVES AND PAINT

The wood most commonly used to make beekeeping equipment is pine. In most parts of the country pine is readily available and cheaper than most woods. Pine sands, planes, and nails easier than most woods and has a good life. Its chief disadvantage is that it rots rapidly when it comes into contact with the ground, but when treated with a good wood preservative, a pine bottomboard will have a life of twenty-five years. Beekeepers rarely treat supers or other hive parts with a wood preservative. If a wooden hive stand is used, it too must be treated.

[The most practical and cheapest wood preservative for a beekeeper to use is a solution of pentachlorophenol; however, this chemical is effective only if the wood is soaked in it for twenty-four hours and is thoroughly penetrated by it. The proper concentration contains 5 percent of the active ingredient.] This concentration is controlled by state legislation and varies from one state to another. In New York State, for example, one may buy a stronger than 5 percent solution only if one

A new bottomboard being soaked in wood preservative in a "boat" made of polyethylene. Without the protection of a wood preservative a bottomboard has a short life.

has a special certificate. The last time I checked it was possible to buy a 50 percent solution in South Carolina (and I presume some other states). The more concentrated solutions are diluted to 5 percent with kerosene. It is, of course, cheaper to buy a concentrate and to dilute it than it is to buy a prepackaged diluted solution. The reason New York State insists that only diluted solutions be sold is that they are safer to handle. Ten years ago 50 percent concentrates were available in New York.

In the tests I have made, for five standard factory-made pine bottomboards piled in a "boat" made with polyethylene, which are rotated each twenty-four hours over five days, one needs three to four quarts of a 5 percent solution to get good penetration; any lesser amount will not give good protection. Painting a 5 percent solution of pentachlorophenol onto wood is a worthless exercise; the wood must be soaked in it.

Pentachlorophenol is not toxic to honeybees, but kerosene and many of the other solvents used are. Bottomboards cold soaked in a pentachlorophenol solution must be air-dried for four to six months to let the kerosene evaporate. Wood air-dries best in the spring and summer months, after which it may be painted or used without painting.

There exist several other wood preservatives, but all are much more expensive, unless one can find a sale, or they leave a greasy surface. Creosote is an example of the latter. It is an excellent wood preservative, better than pentachlorophenol, even when two identical pieces of wood are pressure treated; however, it remains messy and will rub off onto one's clothing for years.

Most beekeepers paint their hives, and there is no question that painting adds to the life of wooden equipment. I have known of beekeepers who have dipped bottomboards and supers in paint thinking it was faster than brushing or spraying. Paint on the inside of supers does no harm, though I doubt if it does any good either.

In the north, it does not matter what color paint is used. In fact, some beekeepers use a dark color, thinking it helps to hide colonies and perhaps protect them from vandalism. In the southern states it is probably advisable to use white or aluminum paint to reflect the heat and help keep the hives cool.

DRIFTING

Drifting is the movement of bees from their colony to another nearby colony. Drifting takes place because bees become confused, not because there is any purposeful movement from one colony to another. In fact, each colony of honeybees has a distinctive odor and great efforts are made by the bees to keep their nest mates together. An apiary is an artificial situation as far as honeybees are concerned. In nature, colonies occupy hollow trees and caves, often great distances from other bees, and drifting is not a problem. Drifting is one means by which disease may be spread from one colony to another and for this reason alone it is worth preventing. Drifting in an apiary is never completely eliminated, but it can be reduced significantly.

The easiest way to limit drifting is to paint supers different colors so that it is easier for bees to identify their own homes.

Bees see several colors distinctly* and by using different combinations of colors, drifting can be reduced. Painting bottomboards and covers different colors also aids in this regard. In the apiary one may reduce drifting further by using landmarks. Leaving trees between colonies is an aid. Some beekeepers orient colonies in different directions. Making straight rows of hives in an apiary is convenient for the operator, but the data I have show that bees are more inclined in this case to drift from one colony to another.

DRESSING FOR THE APIARY

[Stinging insects are much less inclined to attack persons dressed in light-colored, smooth-finished clothing. I have no good explanation for this behavior other than that dark, rough clothing, especially leather, suede, and wool, may have an exterior texture similar to that of animals, such as bears and skunks, that may attack them. Beekeepers prefer to dress in khaki or white cotton clothing and zippers are preferred over buttons. White socks should be used over those that are dark; boots into which pants can be tucked or over which pants may be tied are still better.]

Many types of veils are available. Those made of wire cloth are superior because they are less likely to tear. The portion of the veil through which the beekeeper looks should be painted black, simply because one can see better through a black veil. The personnel who buy veils for the army, to protect against mosquitoes and other biting insects, have not learned this fact and continue to buy difficult-to-see-through veils of an olive, drab color. These are often sold as surplus for bee veils and should be avoided.

Beekeepers should <u>avoid wearing wristwatches</u> because they accumulate sweat and the odor incites bees to attack the

---

*According to Professor Karl von Frisch, honeybees see blue, blue green, and yellow distinctly; red is not visible to them and they confuse it with black. However, on the other end of the spectrum they can see ultraviolet light, which is not visible to humans.

Beekeeper in Florida wearing khaki clothing. Bees are much less inclined to sting clothing that is light colored and of a smooth finish. When one is inspecting the colony, one frame is first removed and placed outside the colony so as to make more room to manipulate the remaining frames.

vicinity of the band. Even metal wristwatch bands will accumulate a bad odor. It is interesting that honeybees and probably other stinging and biting insects are more inclined to attack persons with a foul odor; my only explanation is that they associate such odors with animals that attack them.

Not wearing gloves to work bees is a subject that others and myself have talked and written about at length. Still, they give many beekeepers much comfort. Even commercial beekeepers sometimes find gloves useful, especially when they are removing honey after a honeyflow and the bees are more prone to attack. The chief problem with gloves, especially those made with leather, is that they acquire an odor that may irritate bees in a colony. Rubber gloves may be used and washed periodically, but they are not comfortable to wear.

However, more important than worrying about how to dress, I suggest that one should first open colonies in the presence of an experienced beekeeper and learn how to use a smoker properly, as well as what to look for to keep a colony under control. Guard bees, those of an attack age, are easy to spot once one knows what to look for and a little smoke properly applied will discourage them. One must understand that merely being dressed properly is not full protection against angry bees. There is no question that it is helpful, but one must also learn that bees in colonies inspected at a time when nectar is available in quantity will be much gentler than when there is a dearth. It is best to inspect colonies at a time when the sun is shining and bees, especially the older ones, are flying.

SMOKERS AND SMOKER FUEL

There is no substitute for a smoker when manipulating a colony of bees. I do not know precisely what smoke does to bees, but it probably fouls their sensory receptors and they can no longer detect that their hive is being invaded. Smoked bees gorge themselves with honey, which calms them. It is not necessary to use a great deal of smoke to keep a colony under control, as long as the smoke is applied before the bees attack.

The greatest number of guard bees will be around the entrance, so smoke is applied there first. Smoke is next applied across the frame top bars as the cover is removed. When making the examination, one should remain alert to the behavior of the bees. Whenever a row of ten, twenty, or more heads appears along the edge of a frame top bar, it is time to apply more smoke. Some colonies will need to be smoked before the removal of each frame and other colonies will need only a single smoking during the entire examination.

Smokers are sold in varying sizes. I suggest buying only the largest one available. The better smokers have an asbestos-lined shield around the outside that protects the beekeeper from being burned; this is an added bit of protection that is definitely worthwhile.

All smoker fuels have bad odors. I am frequently asked whether there is not some type of fuel that would leave a less offensive odor on one's clothing. I have never found any such material. Each beekeeper has his favorite smoker fuel. One of the chief concerns is to find something that will stay lit and produce smoke for a long while.

A favorite fuel among those with friends who own cows or horses is used hay-baling twine. Some beekeepers pick, store, and dry such things as corn cobs and sumac cones. Shavings and small blocks of wood work very well and give off smoke over a long period of time. I know one beekeeper who saves the scrap wood from making hive parts and cuts it into one-inch cubes; these chips burn for a long time and the smoker does not need to be refueled so often. It is important to refill the smoker before all the material in it is burned.

WHAT TO DO WHEN STUNG

Once a bee has decided to sting, it moves so rapidly there is little one can do to prevent being stung. Some reaction to a sting is difficult to avoid. Its severity depends in large part upon the age of the bee. Old bees and young bees usually have much less venom than bees about three weeks of age. At this age bees are especially well equipped to be guards because of their temperament, agility, and strength and the quantity of venom they carry.

Once one is stung it is important to remove the sting as rapidly as possible. How one does so does not make much difference: It may be rubbed, squeezed, or scraped from the wound. There has been much humbug published stating that care must be exercised when a sting is removed lest one squeeze more venom from the venom glands and inject it into the wound. The fact is the venom all flows from the glands just as soon as the sting is inserted. Recently I saw a very nice but useless picture in a new book for hobby beekeepers that illustrated the careful removal of a sting with a penknife.

If one watches a sting, one will note that it continues to vibrate for several minutes after it is inserted and the bee has departed; I once observed a sting that pulsated for twenty-five minutes, an unusually long period of time. While the sting is moving in this manner, the sting shafts are being driven deeper into the wound. This will allow the venom to penetrate a greater distance into the flesh, which is to be avoided.

The best treatment that I know for a sting is an ice cube applied to the wound surface. This may slow down the distribution of the venom, but, more importantly, will create a different sensation. It will not stop the pain and swelling, but it is an effective, if temporary, treatment even hours later should there be continuing irritation.

THE BEES IN FEBRUARY

The situation within the beehive in the north in February is not too different from that in January. The data we have indicate that all normal colonies will be rearing brood in February. Food consumption picks up markedly in February, and the danger of starvation is greatest during this month and March. Honeybees have no method of rationing food nor do they seem to be able to recognize an impending food shortage. There is nothing one can do to aid a starving colony during this month in the north. In Maryland and farther south, feeding in February is possible.

Despite the fact that February is the coldest month of the year, I cannot remember a year when the bees did not get at least one flight during the month. The greater the number of brown fecal spots on the snow, and the greater the number of dead bees scattered many feet from the hive entrance, the better the flight. The farther north, the greater the value of feeding sugar syrup in the fall to reduce the amount of fecal matter the bees may accumulate.

I like to visit the apiary at least once during February to check for plugged entrances; also, a winter pack may occasionally be disturbed by a marauding animal or boy. If the cover is removed from a colony, or a pack disturbed so that rain or snow enters the top of the colony, the colony can die. If a colony has been disturbed, one makes the best repair possible and hopes the colony will survive until late March or April, when it can be unpacked and examined more closely.

3
MARCH
Sources of Information

Beekeepers are everywhere and so are sources of information about how to keep bees. Much of the information about bees and beekeeping is free. One may accumulate a wealth of information from state and federal offices and beekeepers' organizations. I suggest that building a small beekeeping library is worthwhile, and winter is a good time to do some reading about beekeeping.

One must be cautious about using techniques and methods that work in one part of the country or world in one's own area without a thorough testing. For example, beekeeping in the Canadian prairie provinces is different from keeping bees in the northeastern states, and the methods beekeepers use in the Canadian west will just not work in New York or New England. In many parts of the hot, dry southeast beekeepers buy and feed large amounts of pollen to stimulate their bees in the spring. They often write glowing reports about their success. One does not need stimulative pollen feeding in the north, yet I am aware that many beekeepers have fed pollen to their bees after reading about the success some beekeepers have had in other parts of the country. It is always best to check with a local source of information before adopting a new practice.

LOCAL BEEKEEPERS

Over 200,000 people in the United States own one or more hives of honeybees. Obviously, not all of these people thoroughly understand bee biology and management, but the beginner should have little difficulty finding a beekeeper with a good background. If one can find a person with ten or more years of experience, especially one who will allow a beginner to go to his apiary, much time, effort, and probably money will be saved.

It is especially important that a new beekeeper receive advice and aid the first few times he makes a colony examination. Honeybees have a good defense system, but with proper timing, care, and the use of smoke, a colony can be thoroughly examined without excessive stinging and sometimes without any stings at all.

THE STATE COLLEGES

Each state has an agricultural college supported jointly by state and federal monies. All of these colleges have three functions: teaching, research, and extension. The system was established by Congress over one hundred years ago, though the extension arm was not formed until the early part of this century.

All state colleges have at least one extension entomologist and about half have an extension apiculturist. These people are responsible for the preparation of literature, usually bulletins, circulars, and mimeographs, in their subject area; they will also answer letters and often arrange and speak at meetings on their specialty. This information is usually free, though recently, owing to increasing costs, charges are being made for some types of literature. However, the chief job of the extension specialists is to provide information to the county agricultural agents; each county in the state has a county agent who is usually housed in an office with the local 4-H agent and federal nutrition personnel. Since apiculture is a specialized aspect of agriculture, not all county agents will know about bees and beekeeping; however, they will know about bee clubs and so forth, either in their county or nearby.

Most state colleges will send one copy of a piece of literature to an out-of-state person without charge. For example, only Pennsylvania and Illinois have circulars on how to build a solar wax extractor; they have been willing to send these to persons from other states, which saves extension specialists the trouble of preparing similar items for their own states.

A beekeeper can build a good library and learn much about the field by writing several states and asking for the available literature. Each year in April, *Gleanings in Bee Culture*, a national trade journal, lists the persons involved with teaching, research, and extension at the state colleges.

APIARY INSPECTION

About half of the states, usually those where large numbers of colonies are rented for pollination, have a state apiary inspector in their department of agriculture. The chief concern of this individual is the control of bee diseases, especially American foulbrood. Some states require beekeepers to register colonies and in some cases their locations, though experience has shown that registration is not required for effective disease control. Most states require that a certificate of inspection accompany bees and beekeeping equipment being moved or sold. Copies of state apiary laws are usually available from the Department of Agriculture at the state capital.

The Federal Government cooperates with state agencies and offers free disease diagnostic service (see the discussion of bee diseases in Chapter 12 for details).

ASSOCIATIONS AND THEIR NEWSLETTERS

As far as I am aware, every state has a beekeepers' organization, and there are many local, country, and regional associations. In addition, the Eastern Apicultural Society is located in the northeast; the Western Apicultural Society is active in the west and northwest. Both of these organizations include the neighboring Canadian provinces. In the southeast there is the Southern States Beekeepers' Organization. These three organizations meet annually, as do the two national organizations, The American Beekeeping Federation and the American Honey Producers Association.

Many of these organizations have newsletters, some monthly and others less frequently. Most of these groups offer special courses for beekeepers in addition to meetings where speakers explore a variety of topics. Information about the local organizations may be obtained by contacting the local county agent, the extension apiculturist or entomologist at the state college, or one of the trade papers or journals.

BOOKS AND JOURNALS

More has been written about the honeybee than about any other insect and most other animals. There is a great variety of books, from romantic novels to good, practical, how-to-do-it guides. There have been relatively few changes in basic management practices in the past fifty years; the swarm control methods used today were described about eighty to one hundred years ago. Views on the best methods of wintering bees have changed, but many of the techniques, even cellar wintering, have merit today and are worth exploring in certain parts of the country. Many of the ideas written decades ago are worth knowing and may still have application.

Spending a little time reading each winter can do much to improve one's management the following spring and summer. Securing a maximum crop of honey is no accident and requires an in-depth knowledge of honeybees. There are very few secondhand book dealers who do any volume of business in bee books; however, I have seen a great variety of instructive bee books in secondhand stores and at auctions. See the Selected Bibliography for a few recommended titles as well as related journals.

CORRESPONDENCE COURSES

Many states, including New York, Pennsylvania, and Ohio, offer correspondence courses in beekeeping; state residence is not required. Most of the correspondence courses do not even require that one own bees. They are intended to guide one through one or more texts and other written material pointing out the ideas and methods of special importance. There is normally a fee for these courses and one is often required to buy one or more books. A certificate is usually awarded to those who complete the course satisfactorily. One does not become an expert beekeeper as a result of taking a correspondence course, but it is a good place to learn the rudiments of beekeeping.

SHORT COURSES

Short courses in beekeeping have become increasingly popular in recent years. Most national and regional beekeepers' organizations offer some kind of short course at the time of their annual meeting. In addition, several states offer courses, including those meeting over a series of evenings, on weekends, and even week-long courses.

A major problem that has become increasingly obvious to those of us teaching short courses is that beginners have little opportunity to open a beehive with some supervision. If one cannot find a local beekeeper who will assist in this regard, then certainly a good place to start in beekeeping is by taking a short course. Information about these is again available from county agents and the state colleges. In addition, the trade journals list most of the events in detail, giving one a wide variety from which to select.

CATALOGS

Nearly all of the major bee supply companies distribute free catalogs on request. Obtaining catalogs from several firms will allow one to compare prices and to learn the types of equipment available. Generally speaking, it has been my observation that the higher priced equipment is of better quality; still, some of the small firms that are trying to obtain a portion of the market often have good beeware.

In my experience, it is better to buy bee equipment from a company that specializes in its manufacture. The large catalog houses that sell a great variety of items often buy their bee equipment on bid and in recent years the quality has not been very high.

Catalogs are helpful in determining what one might buy; however, attending a large beekeepers' meeting where various pieces of equipment are on display can often be more helpful. Before buying a lot of equipment it is advisable to consult other beekeepers in the area.

THE BEES IN MARCH

In the north, colony populations are at their lowest in early March; in southern states, the colony populations will probably reach a peak in March and the orange honey crop may be made. It is important that a beekeeper keep notes about bee activity in his area so as to know when and when not to make colony manipulations.

My notes show that in the Ithaca, New York, area the first bees bring in pollen between March 20 and 25. It is usually between March 10 and 20 that I get several phone calls asking why there are honeybees molesting birds at bird feeders. The bees are not, of course, actually molesting the birds, but they may be at the bird feeders in great numbers on a warm day. Their threshold of acceptance is so low, and they are so desperate to collect pollen, that they will collect anything of the correct particle size, even grain dust if pollen cannot be found. Similarly, in desert areas bees have been known to collect the fine dust ground from dirt roads under dearth conditions. They may also collect coal dust and even sawdust, especially from hardwood trees that contain a little sugar. As soon as it is warm enough for the first plant to bloom the bees will no longer visit bird feeders or collect pollen from anything but real flowers.

On one of the last days in March there will be a day when

pollen is available in abundance and flight will be strong. Such a day is a good time to visit the apiaries and pick up dead colonies. March, in central New York, is still too early to unpack colonies or make any manipulations; however, one can check colony entrances, repair or replace windblown packs or covers, and make sure that the bees are secure.

An elevation of a few hundred feet will make a great difference in colony growth. The apiary on the Cornell campus is at an elevation of about nine hundred feet and the one at the Dyce Laboratory is about two hundred fifty feet higher. One can see a difference between colonies at the two locations, with those at the lower elevation growing more rapidly. The position of those near the campus is also enhanced by the fact that there are many nearby houses inhabited by people who like to grow early spring flowers, which provide the bees with a great deal of pollen.

Beekeepers do not plant honey or pollen plants for their bees, as it is not a paying proposition; however, they may seek out locations near towns and villages where there might be an abundance of crocuses, ornamental pussy willow, or other early spring flowers. Those planted by one homeowner would not provide much forage, but in an area where flower gardening is popular the activities of several persons might make a difference. I once knew a commercial beekeeper who lived in Orlando, Florida, and who kept several hundred colonies within the city limits, where they built up quickly in the spring prior to the orange honeyflow. He told me that the northerners who wintered there nurtured early spring flowers in profusion, much to his benefit.

In rural areas early spring flowers such as skunk cabbage and native pussy willow, common in northern swamps, serve the same purpose. A beekeeper with several thousand colonies is often forced to relocate apiaries, frequently on short notice, because of new buildings, changing agriculture, or road construction. There are many things to consider when selecting an apiary site; while choosing a location near an abundance of one or more major honey plants is the chief concern, a close second is sufficient early pollen, and, to a lesser extent, nectar plants to build colony populations.

4
APRIL
The Start of the Active Season

My notebooks show that about April 1 there is one day or more when bees gather a great quantity of pollen. Such days gladden the heart of every beekeeper. I do not know how much pollen may be gathered, but when I can count ten to twenty bees per minute laden with pollen landing at a colony entrance, I know it is a good day. It is estimated that a cell of pollen and a cell of honey are required to grow a worker bee. Crocuses are one excellent source of pollen; I am especially conscious of them because my neighbor plants them in great numbers. However, I am always surprised at the great number of different-colored pollens bees collect in the early spring, each representing a different plant species.

April is the month when the management of honeybee colonies begins in earnest in central New York. It is important to emphasize that the frost-free date in this area is May 20 and the swarming season starts May 15. Swarming is not a concern in April; at this time beekeepers want to build populations as rapidly as possible.

Two notions should be dealt with at this point. First, I do not advocate or feel one needs pollen substitutes or supplements* in the northern states, especially the northeastern

---

*A pollen *supplement* contains some pollen that serves to attract bees to consume what might otherwise be unpalatable. A pollen *substitute*, which may

states. The use of pollen supplements and substitutes is rare in the north; many have been tried, but they have not become part of beekeepers' routine. In certain southern states, where building populations for package bee or queen production is the chief concern, feeding bees such materials may be necessary.

Likewise, I do not recommend two-queen colonies for honey production.* There are no clear data to support the notion that a colony with two queens will produce more honey than two single-queen colonies. In questions of this nature my position is determined by observing what commercial beekeepers do. Only rarely have I heard of commercial beekeepers using the two-queen management system, though I am aware that most of them have tried it at one time or another.

UNPACKING COLONIES

About mid-April, colonies in the central New York area should be unpacked. Unpacking should be done earlier only if there are an unusual number of dead bees on the ground in front of a colony or on the bottomboard, or if one has reason to believe the colonies may be short of food. Weak colonies should be united and those short of food should be fed sugar syrup. How this is done is discussed in detail later.

In areas as far north as most of New York State (not including Long Island) I believe colonies of honeybees should be given some special winter protection. I wrap my colonies with a lightweight black building paper and place a bun of wheat straw over the inner cover. The telescope cover is stored for winter and the same type of black paper is used as a

---

contain much the same materials as a supplement, has no pollen, though both may use sugar syrup or honey as part of the formulation, again as attractants. Honeybees much prefer natural sources of food over artificial ones, even though the artificial food may be easier to gather.

*The best publication on a two-queen system of management is *Two-Queen System of Honey Bee Colony Management,* Production Research Report 161. It is available for a small fee from the Superintendent of Documents, U.S. Government Printing Office, Washington, D.C. 20402.

cover. The winter pack is designed to provide good ventilation to rid the colony of excess moisture, and to warm up the colony on sunny days so that bees may have a cleansing flight. The pack is not designed to protect the colony against cold. In states south of New York special winter protection and wrapping is probably not useful, though the selection of a proper winter location and the provision of good ventilation are both critical.

In April, beekeepers are concerned with removing any winter packing that might exist, quickly examining the colony for food, and making notes about the suitability of the location. This first inspection is usually too early to check live colonies for disease or the condition of their queen; this comes later, usually in two or three weeks. However, <u>one should make certain that a mature colony in two supers has at least twenty pounds of honey at this time of year. If it does not, the colony should be fed immediately.</u> Bees in colonies with sufficient honey or sugar syrup will devote their efforts to gathering pollen, which is badly needed at this time.

The inner cover, if one is used, or the packing or underside of whatever cover is used should be dry on this first inspection. A damp colony will have suffered a greater loss of adult bees than necessary and spring brood rearing will have been delayed. If the top of the hive is too wet, then either there was insufficient ventilation or the location is one where there is too little air movement. One does not assess the value of a location after one season, but if bad conditions persist year after year, especially after providing the colony with proper ventilation, then one should seek another location. If the inner cover is wet, it should be replaced with a dry one.

It is even worse to find moldy combs in a colony in the spring. One cannot replace moldy combs in an active colony, but making other hive parts dry will aid the problem. Combs in dead colonies will almost invariably be moldy and should be inspected to determine why the colony died. American foulbrood is the greatest danger. The combs, no matter how moldy and dirty, can be salvaged if the colony died of something other than American foulbrood. Dead bees should be brushed off the comb surface gently so as not to break it, and the supers of combs placed in a warm building to dry them and prevent further mold growth. One should not bother to remove dead bees in the cells, since the bees can do that more efficiently later. These supers are given back to other colonies of bees in late May or June, when the colonies are strong, and they will be cleaned up. It is remarkable what bees are able to do with such comb.

In apiaries where the grass is not routinely mowed, I like to put the winter packing paper under the hives (or hive stands) and on the grass in front of the hives. It is a nuisance to inspect colonies with grass growing around and between them and grass blocking flight to and from colony entrances is a problem for the bees. A few stones will hold the paper in place.

UNITING WEAK COLONIES

I define a weak colony as one with less than about a pound and a half of bees, or about six thousand bees. It is difficult to estimate this number of bees, and perhaps a better measure is

Grass blocking a colony entrance. This is a nuisance, and a piece of cardboard or the leftover winter packing paper under and in front of the hive will prevent the situation.

the amount of brood such a unit will have. In this case a weak colony would be one with patches of brood on the two sides of one comb but on only one side of a second comb, and perhaps less. It is important to remember that to rear brood a colony must keep the brood nest (brood rearing area) at 92°F or above. This does not mean the whole interior of the hive body is at this temperature—only the portion with eggs, larvae, and pupae in brood cells. Bees control the brood nest temperature by forming a cluster around the brood. One worries when there is a small brood nest, especially when there is a patch of brood on only one side of a comb in such a nest.

It is logical to expect two to four weak colonies in an apiary of thirty to forty colonies. If there is a greater number, and the colonies were apparently in good condition in the fall, then it may be time to make a critical examination of the apiary site.

Weak colonies in April may survive the spring, but they will not become producing units. To explain this last statement better, one might think of a three-pound package in mid-April. My experience in the northeast is that a package will not produce a surplus the first year, for its population is not great enough. Even though weak colonies may have the advantage of some brood, this is still not sufficient. Most weak colonies I have seen in April have the added disadvantage of having some moldy or wet combs.

In April, weak colonies should be united. This is done by taking from each hive the super that contains the brood and placing one above the other (I usually place the weaker of the two units above the stronger one), with a single sheet of newspaper between the two. A few slits in the newspaper cut with a hive tool will aid the bees in removing the paper. If the bottomboard is wet, a dry one should be used. An entrance cleat should be positioned so that five to ten bees can exit at one time. It is not necessary to kill one of the two queens, though I have noted that most experienced beekeepers will remove the queen from the weaker of the two colonies. The new hive should have a minimum of twenty pounds of honey.

Two and sometimes three weak colonies combined in this

manner will grow and should produce some surplus honey the same year. It is even possible to have a sufficient number of bees to make a nucleus colony from the unit in June or July.

Combining weak colonies in April in the north reduces one's colony numbers and might not appear to be an economical move; however, considering the stress placed on growing colonies by the wide fluctuations in temperature that occur in the spring, the move is logical.

THE FIRST INSPECTION—DISEASES

The primary disease with which beekeepers are concerned is American foulbrood. Control of this disease is under the supervision of a state apiary inspector in some states, and in others nothing is done to prevent its spread. American foulbrood is caused by a bacterium that infects and kills bees in the larval and pupal stages. It has no effect on humans or any other animal. New York State has an effective inspection system that has kept the incidence of disease near 1 percent for many decades. Infected colonies found by inspectors are burned or fumigated in accordance with state rules and regulations.

I have never advocated the use of drugs to control American foulbrood in New York, though drugs are widely used in the state and elsewhere. In states where there are rigid disease control programs drugs are not needed, but where there is no inspection it is not unusual to find 10 percent of the colonies infected and losses high.

The use of drugs to treat diseases is under the control of the Pure Food and Drug Administration (FDA), which determines when a drug may be purchased over the counter with no controls by anyone. Drugs for human and most animal use must be prescribed by a physician or veterinarian. However, in the case of some animal drugs, including those used to treat bee diseases, the FDA states that if ample information is available to the layman to use the drug intelligently and avoid misuse, then no control is needed. In the background is the FDA's awareness that there are not enough veterinarians knowledgeable about bee diseases to inspect and control the problem. This same thinking is behind the lack of control on drugs to treat poultry and some other animals. The result is that the use of drugs to control American foulbrood is unregulated and widespread.

I am also aware that many beekeepers routinely treat colonies whether they need it or not. The FDA is opposed to preventive feeding in some instances. In the case of bees, FDA states that such feeding is not preventive feeding. The feeding of drugs to cause an animal (such as a beef cow or pig) to gain more than normal weight is prohibited; however, the periodic feeding of a drug to bees is not considered preventive feeding. Preventive feeding must be routine and done every day to fit FDA's definition. I do not defend, nor do I like, this reasoning, but am here merely reporting the official position. The beekeeper is forced to determine the situation in his state and to decide what course of action to follow. Drugs for American foulbrood are readily available and are usually sold with adequate directions. Fortunately, there has been no drug-contaminated honey on the market, nor is such a problem likely to be encountered if directions are followed; however, that is not the question addressed here.

Any other diseases that are present in a colony, such as European foulbrood, sacbrood, or nosema disease, are what are called stress diseases. They are present when colonies are stressed as a result of poor location or inadequate food. In my opinion these diseases can be controlled by the beekeeper (see the section Selecting the Apiary Site in Chapter 1). More discussion of bee diseases is given in Chapter 12.

THE FIRST INSPECTION— FOOD AND FEEDING

At the time of the first April inspection a colony in the north should have at least twenty pounds of food in reserve. Frames may be removed from colonies with twenty-five or more pounds, and given to those with less food. Estimating food reserves is not too difficult; a full-depth frame of honey,

Three one-gallon glass jars used to feed a colony. The jars are inverted directly over the top bars; any burr comb on the top bars is first removed. When feeding colonies one should never use less than one gallon of syrup, and more than one gallon is preferable.

filled and capped (sealed with wax), will contain five to seven pounds of honey if the spacing of the combs is nine per super. Fatter combs may contain another pound or so of honey.

It is worth repeating that foragers from colonies with sufficient honey will devote their attention to pollen collection. Honeybees appear to know when they have sufficient honey. I wish I could understand how they determine this.

Colonies with too little food should be fed without delay, regardless of the weather.

If a colony needs feeding, one should never feed less than a gallon of sugar syrup; I prefer to give two or three gallons at one time. The feeding should be done so that the bees will take the food from the feeder quickly, ripen it properly, and store it as soon as possible. Under ideal circumstances feeders will be empty in two to five days. This is true with both spring and fall feeding. If a longer time is required, there may be mold growth in the feeder. Mold usually does no great harm, though it may make the feed unpalatable; in rare circumstances I suspect some toxins may be produced that could have an adverse effect on the bees.

[In the spring, sugar syrup should be made with one part sugar and one part water by weight or volume; in the fall I use two parts sugar and one part water. I am convinced bees will start to take syrup faster if it is warm (90° to 100°F); it is also easier to get the sugar into solution if the water is warm.]

I use one-gallon, wide-mouth, mayonnaise-type glass jars for feeding, punching about thirty holes in the cap with a threepenny nail or making a similar number of holes with a $1/16$-inch drill bit. Drilled holes are a nuisance to make, but easy to clean. I am not enthusiastic about glass feeders, but know of no better way to rapidly feed bees the volume they need. Division board feeders would be my second choice, and I have observed that several commercial beekeepers use them. A division board feeder is a box with an open top the size and shape of a frame, which may be placed in a hive in place of one or two frames and filled with sugar syrup. Because it is close to the brood nest the bees will be able to feed from it day and night. At one time most beekeepers used tinned ten-pound pails, but too little tin is used today on pails and they rust too fast. Miller-type feeders, pans, or wooden trays that fit into half-depth supers are favored by a few beekeepers, but there are mold problems with those made of wood, rusting with those made from galvanized iron, a short life with those made of plastic, and other unfavorable reactions with aluminum pails, which result in the formation of pinholes that cause them to leak.

Entrance- or Boardman-type feeders should be avoided in the north; they were designed for use in the southern states for feeding nucleus colonies and hold too little feed. Mold growth in the sugar syrup they may contain is encouraged if these feeders are exposed to sunlight. They are too far away from the clustered bees at night or when it is cool; I have observed bees in colonies with entrance-type feeders starve to death when the food was only six inches away. Entrance-type feeders are popular because one can see them and observe the bees taking food. I do not understand why the manufacturers of beekeeping equipment continue to advocate their use without stating the conditions under which they should and should not be used.

A wooden carrying box made to transport feeding jars and protect them from breaking.

I have built special wooden boxes that hold four one-gallon glass jars to carry the jars to and from the apiaries. The boxes have dividers that prevent the jars from banging against one another. Building the boxes takes time, but without a good box to hold them the jars would break.

When we note that a colony needs feeding we remove the cover and inner cover and give the bees a good smoking (three to six strong puffs). In a populous colony a second smoking may be necessary after five to twenty seconds. The purpose of the heavy smoking is to drive the bees down off the top bars of the frames in the upper super. I then scrape the top bars to remove the burr comb and propolis, obtaining a level surface on which to place the feeder jar(s). As mentioned earlier, each jar cap should have thirty to forty holes from which the bees may suck the syrup; this number of holes is needed, since many will be blocked by the frame top bars when the jars are inverted.

An empty super is placed around the jars and the cover and inner cover put on top. Sometimes it is necessary to invert the inner cover so that it does not rest on top of the jars. Many years ago beekeepers placed burlap bags around the jars to help retain the warmth and to deter comb building around the feeders, but now it is almost impossible to obtain burlap. If the feeders are left in place for a short time only, the bees will not have time to build burr comb around them. If the bees do not take the food promptly, within a few days, the colony might not be normal, perhaps not queenright.

THE FIRST INSPECTION—BROOD PATTERNS

Brood is the name given to the developing eggs, larvae, and pupae. Since bees maintain a constant brood rearing temperature, and control the humidity, though to a somewhat lesser extent, it is important that they keep the brood nest, or rearing area, compact. In a developing brood rearing area eggs are adjacent to eggs, larvae to larvae, and pupae to pupae; presumably this makes feeding in the larval stage easier and the nurse bees are drawn to those areas where there are great numbers of larvae.

One does not expect that every cell in the brood nest will be filled; preferably 90 percent or more should be filled. Cells may be empty for a variety of reasons, but when empty cells appear in large numbers, the colony should be checked first for disease and then for starvation. If these are ruled out, the blame is put on the queen and she is replaced. Queens can fail for a great variety of reasons, including disease, improper mating, old age, poor nutrition, physical handicaps, and so on, and it is often difficult or even impossible to determine which of these is the problem. One frequently replaces a queen without knowing precisely why she is failing. Commercial beekeepers often make notes about the brood patterns of queens from different breeders, in addition to keeping a careful record of production. I think an outdoor bee club meeting devoted to studying and assessing brood patterns would be a worthwhile project.

EQUALIZING COLONIES

Colony populations will vary greatly in an apiary in the spring. Some colonies will be so populous that they may swarm, while others will be too weak to build up properly. I have already stated that very weak colonies should be united; however, moderately weak colonies may gain in strength and strong colonies may be weakened by exchanging their positions in the apiary. This is best done when there is good flight to and from the colonies and the bees are actively foraging. If properly done, the weak colonies will pick up much of the foraging force from the strong colonies and the populous ones will lose bees, making swarming less likely.

There is always some danger of fighting among the bees in the weaker unit when equalizing is done in this manner; however, much of the potential for fighting is eliminated by exchanging positions when there is active foraging. My observation is that bees laden with food coming into a hive are rarely challenged when they go into the wrong hive. The odors of food (from the flowers visited) will often offset bee recognition, which is usually mediated by what is called hive odor, a composite of all the odors in a hive.

A good time to exchange colonies is when the first thorough spring inspection is made in late April. The colonies are usually lighter than they will be any time later in the year, making the physical work much easier. There is certainly less likelihood of fighting than when the colony populations are greater.

[The greatest danger in equalizing is the possibility of spreading disease. The chance of doing so can be reduced by conducting the first disease inspection at the same time. Equalizing colonies by exchanging positions is certainly superior to equalizing populations by giving weak colonies frames, or partial frames, of brood, since they may often have too few bees to keep it warm. In most weak colonies the quantity of brood is limited by the number of adult bees available to protect it. When a weak colony is strengthened by adult bees, it will respond by rapidly increasing the size of its brood nest.]

PACKAGE INSTALLATION

[Package bees are bees sold by the pound in wire-bound wooden cages. I advise the use of three-pound packages in the

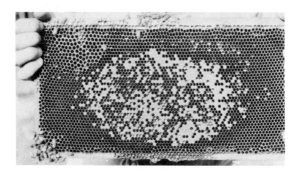

A very poor brood pattern indicating a failing queen or disease. Note that a large number of cells contain no pupae and are not capped (covered). A colony with a brood pattern of this nature must be checked closely.

A good brood pattern with a relatively small number of cells not capped.

An excellent brood pattern. All of the brood is in the pupal stage. A normal full-depth frame contains about 6,800 cells (on both sides) and what is seen here is about 2,500 cells of brood, which will probably emerge at much the same time.

An excellent brood pattern. This picture demonstrates how queens lay eggs in concentric circles. The open area in the center of the comb contains eggs and young larvae; brood emerged from these cells only a few days earlier. Note that in the area where there is capped brood, food cells are empty. In a comb with a good brood pattern, brood of the same age is together.

Raised cappings on the brood cells on a comb of predominantly worker cells. This is an abnormal condition. In this case there are drones being raised in worker cells from eggs deposited either by laying workers or a drone-laying queen (a queen that has exhausted her sperm).

northern states; in areas where there is an intense honeyflow, especially parts of Canada, beekeepers may use two-pound packages. Packages should always be shipped with a queen that is in a small separate cage; the only reason for caging the queen is that one can make certain she is alive when the package arrives in the north.

The ideal time for a package of bees to arrive in the north is about April 15. Packages that arrive later will need more care, especially feeding, in the fall. There may still be snow on the ground at the higher elevations, but spring flowers will soon be in bloom. The bees in the package should be fed within an hour or so after arrival, while they are still in their shipping cage, until they are fully engorged. The bees in most three-pound packages will consume a pint of sugar syrup within an hour or so. The best way to feed syrup to bees in a

Painting sugar syrup (half sugar and half water) onto the screen of a package of bees. When a package of bees is received from the south, they should be fed a pint or more of sugar syrup. The bees will lick it up rapidly.

Removing the metal feeder pail that is shipped with the package. This is done after the bees have been fed, and the pail is then discarded. The queen cage, attached inside the package with wire, is removed and the queen's condition checked.

package is to paint the syrup on the screen with a small brush. If one is installing two or more packages or installing a package in an apiary, the ideal time to put package bees in a hive is during a light rain or in the evening; the rain and/or darkness will reduce drifting. There is no problem if one has only one package and there are no other bees in the vicinity, since the package bees have no other potential home to confuse with their own.

To install* a package of bees one removes the wooden block covering the feeder pail in the package. The pail is removed and discarded. The queen cage is usually tied to a staple near the large hole in the package. It is removed and the

---

*The installation of package bees is treated more thoroughly in illustrated *Package Bees: Their Installation and Immediate Care*, Information Bulletin 7 (8 pages, 1981), by G. F. Combs, Jr., and R. A. Morse, available for a small fee from the Mailing Room, Building 7, Research Park, Cornell University, Ithaca, N.Y. 14853.

queen checked to make certain she is alive. The cork on the candy end of the queen cage is removed; normally I remove some of the candy so the queen will be freed within twelve to twenty-four hours by bees that eat the remainder of it. The package bees are then dumped from the package into the hive. This is done rapidly to reduce the number of bees that might take flight. It is usually necessary to give the package a few good strong shakes to dislodge the bees (one never gets all of the bees shaken free); when only a hundred or so remain, the package is placed in front of the hive, hole up, and the remaining bees allowed to crawl from the package into the hive.

The bees in the package should be fed at least a gallon of sugar syrup immediately after they are installed in a hive; a package of bees will consume up to thirty pounds of sugar in the form of syrup during their first month in the north. One should not let the bees be without feed during the first month. As with mature colonies, feeding sugar syrup allows the bees to concentrate their efforts on pollen collection.

Packages are installed without the use of smoke. In fact, smoking would do little good, since the bees have nothing to feed upon and the chief reason for smoking bees is to cause them to engorge; however, making certain the bees are engorged is an important aspect of package installation. Bees in packages will not sting if properly fed, unless, of course, they are squeezed or caught. However, the first time the package is inspected, smoking and the normal precautions used when opening a colony should be taken.

One reads much about whether package bees should be installed on drawn combs or foundation. Foundation has been preferred because of the fear that the package bees might be carrying honey infected with disease spores. If foundation is used, that honey will be consumed in the process of building new comb and there will be less danger of the new colony becoming infected. Beekeepers worry much less about this today than they did years ago. The package bee shippers are very much aware that the beekeepers to whom they sell their bees want clean bees, and they take strong measures to ensure

A super that is to become some package bees' new home. Four frames have been removed. The bees are shaken into this open space rapidly, the frames replaced, and the hive closed. Sometimes the mass of bees on the bottomboard may prevent the frames from settling into place immediately, but after a few seconds their weight (do not apply pressure to the frame) will force the bees to move and the frames will fall into place.

A scale hive outside my office. Such a hive is both fun and important in determining when a honeyflow is in progress.

the bees are disease free. I much prefer to use drawn combs, since the bees in the package get a faster start, being spared the trouble of preparing the new comb. As in the case of mature colonies, a sunny slope facing east or south is the best location for a new package. The entrance cleat should be kept in place until about mid-May or June 1.

THE SCALE HIVE

A scale hive is a hive on scales that are reasonably accurate and which one can read every day or as desired. I think it would cost too much to buy a new platform scale, and I am not convinced that some of the simple models on the market are adequate or will have a long life. I use a scale that is at least fifty years old, which I found in a secondhand store; it weighs accurately to about a fourth of a pound.

The scale hive indicates when the honeyflow is in progress. If one keeps records over a period of several years it is possible to predict, within reason, when a honeyflow will take place; this allows one to design a management schedule that is not just guesswork. But a scale hive is more than just a practical tool—even experienced beekeepers get excited when a scale hive shows a weight gain of five or ten pounds in a day. The honey I harvest is made in a relatively short period of time; knowing when and how much is being made is satisfying and a testimony to the value of one's work and effort.

5
MAY

May is the most hectic and critical month in most beekeeping operations in the north. The last frost occurs in this month; it is the height of the mud season and getting in and out of beeyards is often difficult. Colony growth is rapid. My notes show that nectar (unripe honey) may run freely* from uncapped cells in the brood nest during the first week of May; the source is primarily dandelions, which are in full flower in protected areas but not yet in full bloom in the open fields. Dandelions, an accidental importation from Europe centuries ago, are almost everywhere and are an important source of both nectar and pollen for bees. My notes indicate that on May 4 of a recent year I gave a thorough inspection to eight colonies in a small apiary with no veil and only a small amount of smoke. The field forces in the colonies were so engrossed with pollen and nectar collection from dandelions that they paid no attention to me. This is not too unusual at this time of year.

About mid-month the first apple blossoms appear. The first dandelions go to seed about May 20 and apples are in full bloom between May 20 and 25. Yellow rocket, a plant that

---

*During inspections, the combs should be held over the exposed top bars of the super so that any dripping nectar falls back into the hive, and not onto the ground, where it might incite robbing by other bees.

resembles mustard, flowers at about the same time as apples. A populous colony of bees may store twenty to forty pounds of surplus honey during May, but it is never harvested. The honey gathered in May is not usually of very good quality, being half dandelion, fairly dark, golden yellow, and containing a great deal of pollen. More important, the bees will need that food during the continuing buildup. Colony populations do not reach their peak until about July 1 and one does not want to do anything to discourage their growth in May.

HOW TO INSPECT A COLONY FOR CROWDEDNESS

There is a simple, rapid way to check colonies in early May to determine if they are congested and when swarm prevention measures should be taken. One need not remove individual frames; in fact, doing so delays the process and can be misleading. At this time of the year the colony should be in two or three supers. To make a quick assessment of the situation one removes the outer cover, leaving the inner cover in place, cracks the supers apart, gives the bottom half of the upper super a couple of strong puffs of smoke to drive the bees off the bottom bars of the frames, tips the upper super forward and upward, and examines the top bars of the lower super and the bottom bars of the frames in the upper super.

In all of beekeeping the most important consideration is being able to assess what is taking place at the time of this examination. A decision should be made in one to five seconds. In an apiary of twenty to sixty colonies, the beekeeper will check four to six colonies at random in this manner. On the basis of what he sees in these colonies he decides to move on to another apiary, to take swarm prevention measures, or, if he is behind or has made a mistake earlier, to undertake swarm control* measures.

---

*Swarm prevention is concerned with those measures the beekeeper takes to prevent the laying of eggs in queen cups. Swarm control is concerned with those measures the beekeeper takes after there are larvae in queen cells and it is apparent the colony is about to swarm.

Checking for congestion. One smokes the colony and pries up the top super, looking for queen cups (and/or queen cells) along the sides of the frame bottom bars.

This colony is not yet congested. One can count about ten queen cups and only a few are enlarged.

At this time of the year bees should occupy both supers. If there are no bees in the lower super, the colony is obviously weak, and if only a portion of the frames in the upper super is occupied by bees, it might even be advisable to remove the lower super, though this action really should have been taken in the April inspection.

Assuming that both supers contain bees, one asks how many frames in the upper super contain brood. How many queen cups are present on the upper super's bottom bars? How congested is the colony? The answers to these questions

A congested five-frame nucleus colony. Here congestion causes the bees to cluster on the outside of the nucleus box.

A crowded colony in danger of overheating. This forces many bees to cluster outside of the hive; only a rain or cool weather will force them inside. The colony needs supering and probably upper ventilation; no doubt queen cells are being constructed and the colony will soon swarm if not given attention.

determine what is to be done next and are the subject of the following two sections.

SWARM PREVENTION

Three recent studies indicate when the swarming season will occur in the areas studied. A six-year study in central New York State shows that 80 percent of swarming occurs between May 15 and July 15, and 20 percent between August 15 and September 15. A two-year study in Maryland indicates that most swarming in that state takes place between April 10 and June 15. A one-year study near Davis, California, indicates that nearly 90 percent of swarming in that area occurs during April and May. State extension apiculturists may have some informa-

Swarm emergence dates, 1971–1976, for the Ithaca, New York, area. For a period of six years my students and I collected all of the swarm data brought to our attention, mostly from phone calls, in the Ithaca area. The result was this graph, which tells us when swarming is most likely to occur in the area.

tion on the subject for other states. Beekeepers need more data for other parts of the country. In my opinion, keeping a record of when swarms are seen and captured in one's area is critical to devising a management schedule.

Four to five weeks before the swarming season starts, queen cups begin to appear on the bottom bars of frames in populous colonies. Normal colonies destroy queen cups in the fall and start new ones in the spring.

During the first week of May in my area I expect to see five to twenty cups hanging from or alongside of the bottom bars on frames in the upper super. This is the time to add a third super of drawn, empty combs to a two-super colony. Single-story colonies should likewise get one super of combs. Very populous colonies, those that have brood in the upper super in seven or eight frames and some brood in the lower super, should be reversed, that is, the lower super placed on top and the top super put on the bottomboard. This splits the brood nest and puts the greatest quantity of brood below, so the queen has room to work upward. I rarely reverse without giving the colony a super of combs as well.

An alternative to reversing is to put a frame with a small amount of brood in it into the newly added third super. This is called spreading the brood and is done only when one is

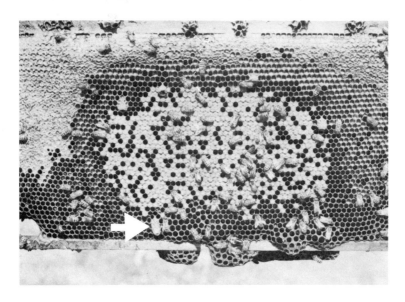

The single queen cell (arrow). The one shown here is a typical emergency queen cell. To make this cell the bees selected a worker larva and enlarged the cell around her.

Two natural swarm cells. Contrast these with the emergency cell. Note that these do not have a worker cell as their base, but are built in the wider-than-normal space between two combs.

A natural queen cell. Queen cells are usually the same color as the wax around them, as they are started from queen cups made of the wax in the vicinity. However, as the larvae grow, the cell is enlarged and here one sees new white wax being added. This is called whitening the cell and indicates that some nectar is available that is stimulating wax gland development.

Two empty queen cells from which queens have emerged, as can be seen by the rough texture of the opening.

certain the colony has sufficient bees to cover the brood and keep it warm. In populous colonies a frame of brood might be elevated in this manner and an empty comb placed in the center of the brood nest below.

All of these measures are designed to relieve congestion and prevent swarming. If they are taken too early, brood may be chilled; if taken too late, the queen may lay in the queen cups and swarming may then be imminent. The number of queen cups a colony builds varies with the race of bees, the time of year, the quantity of food available, congestion, and probably a host of lesser factors. The number of cups is an indication, and not a precise measure, of what might happen.

Measuring the degree of congestion in a colony in May is an art. I think it is obvious that the stimulus to build queen cups has its basis in hive chemistry and relates to congestion. Building queen cups does not disrupt normal colony life, but as soon as bees have larvae in queen cells several changes take place: Worker bees cease to forage or undertake household chores and begin to engorge with honey, worker wax glands begin to develop, queens begin to lose weight in preparation for flight with the swarm, egg laying by the queen is reduced, and scouts turn from food searching to home hunting. All of these processes begin at about the same time, five to seven days before the swarm exits. The swarm may leave the hive any time after the queen cells are capped.

When cups with eggs or one-day-old larvae are present in a colony, the swarming process may be reversed by removing the cups and relieving the congestion. However, once two-day-old larvae are present, most beekeepers agree the process is irreversible, and swarm control measures must be taken, requiring strong action. It is important to do all one can to prevent swarming, but one must be on time. Beekeepers call this keeping ahead of the bees; it is part of successful beekeeping.

If I make an inspection of the preceding nature the first week of May in my area, I like to make a second similar inspection about two weeks later.

SWARM CONTROL

Once queen cells in a colony contain larvae two or more days old, major changes have taken place in the body chemistry of the bees in the hive (see the section on Swarm Prevention). Relieving congestion at this stage will not usually stop the colony from swarming.

To control swarming one must do one of three things: remove the brood, remove the queen, or separate the queen and the brood. (An alternative to removing the queen is to find and cage her to prevent her from further egg laying.) Obviously, any of these practices is time-consuming and certainly impossible on a large scale. Thus one can see why it is imperative to undertake swarm prevention measures on time.

Demareeing, a practice named after beekeeper G. W. Demaree of the late 1800s, is a system whereby the queen is placed into the bottom super with no brood, and the brood is placed three or four supers above. The queen is kept in the bottom super with a queen excluder. The chief problem is that the brood, placed in a super far above the queen, is so far from her that the nurse bees with the brood think they are queenless and build queen cells with the brood. One must go through the brood seven or eight days after making the separation and cut out the queen cells. If one queen cell is missed, the colony will swarm. I have used the Demaree method to correct an error, and it takes at least fifteen minutes to treat a colony in this manner. I can usually find a queen the first time I search for her, but once in a while I miss her, and when I do, I waste much time going through all of the frames again. Under these circumstances Demareeing takes more than fifteen minutes per colony. It is not a practice I recommend, but it is one of those things every beekeeper should try at least once.

I hope these few comments on swarm control emphasize why swarm prevention is so important.

CLIPPING AND MARKING QUEENS

Clipping one or both of a queen's wings with scissors will prevent her from flying and accompanying a swarm. The bees in a congested colony will not realize that their queen cannot fly and will repeatedly try to swarm. A queen will attempt to depart with the swarm each time and there is the possibility that she may succeed in moving a foot or more from the hive and be lost. However, generally, if the queen cannot fly, the bees return to the hive and keep trying to swarm, sometimes one or more times each day. Eventually a virgin queen will emerge and the swarm will depart with her, leaving the old queen behind. I have no precise data on what takes place, but she is probably eventually killed by another virgin that emerges in the hive. While clipping a queen's wings delays swarming, it does not stop it. In the interim, the bees in the colony remain engorged with honey, the scouts continue to inspect new homesites, the queen does not lay, and honey production is halted. Thus it is clear that clipping a queen's wings is neither a swarm prevention nor a swarm control method.

It may be useful to clip a queen's wings to be able to identify her in the future or to determine her age. Placing a paint mark or a plastic disk on her thorax may be equally helpful in this regard. Bits of paint, or a disk, may wear off, though usually such objects are glued firmly both to the thorax itself and to the hairs on it.

To clip a queen's wings one picks her up by the wings with one hand and transfers her to the other hand so that she is grasped by the thumb and forefinger from the underside. Queens rarely attempt to sting. In this position the wings are exposed and can be cut with a pair of fine scissors. I have known of beekeepers who cut off one wing in one year and the other wing the next year. Removing half the wing or cutting the wing on a diagonal may identify her further. Removing part or all of a queen's wing does her no harm. Devices have been invented to hold the queen immobile against a comb while clipping, but in my opinion it is faster and easier to grasp the queen firmly, make the cut, and finish the operation within a few seconds. One may be a bit skeptical about one's ability to grasp a queen without injuring her, but it is neither difficult nor dangerous to the queen. My experience is that queen honeybees, while delicate, are capable of tolerating a great deal of handling. One grasps the queen in the same way to mark her with fingernail polish or paint or to glue a disk to her thorax.

RENTING BEES FOR POLLINATION

Most beekeepers make their living producing honey. I have never seen any accurate figures, but at least 90 percent of beekeepers' incomes across the country comes from the sale of honey and beeswax. The percentage may be even higher. Still, when beekeepers defend their budgets for research, extension, teaching, and disease control, they point to the role the honeybee plays in pollination. Their task, in this regard, is made easier each year as fewer people produce the food for the nation, as fields, orchards, and groves become larger, and as the need for honeybees becomes greater. Many insects act as pollinators, but only the honeybee is easily moved and available for work whenever she is wanted.

Still, if I were a commercial beekeeper, I would not be greatly interested in renting bees for pollination. More money may be made producing honey. Beekeepers who rent bees for pollination face a myriad of problems from pesticide poisoning, to moving colonies during the mud season, to collecting rental fees. Rented bees are exposed to disease, a very real danger in some areas.

The advantage of renting bees for pollination is that it gives a beekeeper some income, usually in the spring when there is little else to be sold from an apiary. In California, beekeepers may rent bees for the pollination of almonds, both to gain a pollination fee and to provide the bees with a much-needed source of pollen to build colony populations. I am told almonds are excellent buildup plants for bees.

In the United States about fifty agricultural crops of economic importance require or are benefited by visits by insects to produce seed or fruit. The importance of these crops varies by state. For example, Maine, Michigan, and New Jersey are concerned about blueberry pollination, whereas in Massachusetts more bees are rented for the pollination of cranberries than any other crop. Throughout the north bees are in demand for apple pollination. In some areas growers have been forced to keep their own bees because there are too few beekeepers to supply the needed bees. No two locations are alike and in each area a beekeeper must decide how he will manage his operation.

A chief problem in many pollination situations is that the bees may be needed when colonies are undergoing rapid growth. This may complicate the question of congestion and swarming; one would not care, for example, to add supers to a colony while the bees were temporarily in an apple orchard, as it would make carrying them out of the orchard more difficult.

METHODS OF MOVING BEES

[Under normal circumstances honeybees forage over an area two to four miles in diameter. If a colony is moved only a short distance, a few feet to a mile, the foragers will return to the old location. However, if a colony is moved several miles or more, the foraging area will be completely new to the bees and they will reorient to their new homes with ease.] One study found pollen-foraging bees returning to their hive in large numbers only twenty minutes after being placed in a new location. This emphasizes what has been said before about the adaptability of the honeybee. With proper care it is a simple task to move colonies.

Moving colonies is discussed here, since one of the few reasons to move colonies in the north is to use them as pollinators. To move a colony one should first make certain that the frames in the supers fit firmly in place and will not move or sway so as to crush bees. In most colonies there is enough burr and brace comb that this is no problem. If one has new equipment or if there is little burr comb, the frames should be

Moving screens. This moving screen is about one and a half inches deep and provides a good clustering space for the bees. It is made of eight-mesh hardware cloth, which is tougher than window screening. The four nail holes are predrilled, which makes it easy to remove the nails that hold the screen in place.

Moving screens that are left in place too long, as this one was, may accumulate some burr comb. This is easily removed with a hive tool and does no harm.

pushed firmly together and a small nail(s) used to hold them in place.

The next task is to nail, staple, or strap the bottomboard and super(s) together. This is a critical task, for if a hive breaks apart en route, one will have many angry bees to contend with. Bee supply manufacturers sell hive staples that work well and which are easily removed, but strips of wood will also do the job.

One of the chief problems in moving bees is to provide them with sufficient ventilation. It is very easy to excite the bees and cause them to lose their ability to ventilate. If this occurs, the temperature in the hive may rise rapidly and the combs may melt and the bees suffocate. Ventilating and keeping bees cool is a major problem for those who move large numbers of colonies.

There are two basic ways to prepare hives for moving. One is to screen the colonies with an entrance screen and a top screen, and the other is to use no screen and to move the colonies with an open entrance. Open-entrance moving is increasingly popular because if the bees become too hot, they can crawl out of the hive. The chief disadvantage in moving open entrance is that a large number of bees are over the hives and being stung a number of times is unavoidable. When open-entrance moving is done, one screens the whole load with wire cloth on a wooden framework or with a heavy plastic screen.

If colonies are individually screened, it is important to use both a top screen and an entrance screen. Top screens are usually built with an inch-and-a-half or two-inch-deep rim, the same size as a standard super; this provides an extra space where the bees may cluster. Several types of entrance screens may be used; the simplest is a piece of ordinary window screening cut two inches wide and fifteen inches long and tucked into the entrance, where it will remain in place if properly cut. The top screens may be put on at any time and the entrances are screened after dark, when the bees have stopped foraging. If one moves open entrance, then the colonies are picked up after dark or early in the morning. Heavy smoking will keep the bees in their hives. If colonies are strong

An entrance screen made of ordinary window screening. When it is pushed into place, the rough wire edges hold it firmly in place, and only rarely will such a screen work loose.

and bees are hanging out of the entrance when one is preparing to move, they can be forced to move inside by smoking.

Beekeepers agree that leaving a truck engine running while loading and unloading colonies appears to calm bees. I presume this is because of the vibration, but there are no good data on the subject. When colonies are moved, many bees will go to work immediately; however, moving also appears to anger bees and often there are many aggressive bees in the vicinity, especially when entrance screens are used. It is just as well not to work or manipulate colonies for a day after they are first moved to a new location.

PESTICIDES

Pesticides are mentioned at this point because the first losses due to their misapplication may appear in this month. Beekeepers have suffered losses from pesticides since about 1880, when the first insecticides—Paris Green and London Purple, compounds containing arsenic—were introduced and widely used. The problem became increasingly acute during the 1930s and especially the 1940s, as certain orchard pests became resistant to arsenic and both the number of sprays and the quantity of arsenic applied each time increased. Some honeybees were killed as a result of drinking pesticide-contaminated water from wheel ruts and other holes in orchards. The arsenicals killed bees slowly and foragers were able to return to the hive with contaminated pollen loads. The first thing a young bee does when she emerges from her cell is engorge on pollen and nectar. The adult field bees died after several trips to fields where lead arsenate was applied, and a few days to several weeks later the young worker bees died. However, in the late 1940s the apple growers using lead arsenate recognized the danger, and without urging from the beekeeping industry they changed their spray practices.

This was also the beginning of the DDT era. There is no record of a single honeybee colony being killed by a normal application of DDT. In fact, even several applications of DDT to a field where honeybees are foraging cause no problem. It appears, for reasons that are still not clear, honeybees can break down the DDT with which they come into contact and it is not toxic to them.

The discovery of DDT prompted many industry, university, and private researchers to seek chemicals that might protect plants and animals against insect pests. As a result, thousands of materials with insecticidal activity have been discovered and hundreds have been manufactured and sold. Lists of the common insecticides, noting their toxicity to honeybees, have been prepared. Researchers in California and Washington State have been especially active in this regard. Beekeepers everywhere in North America refer to these lists, and experience has shown that the data are valid across the continent. Copies of these lists, which are revised frequently, are available from the local state apiculturist or entomologist. Several insecticides are a problem for beekeepers today; two have been especially troublesome and they are discussed below.

In 1959 the Union Carbide Corporation brought a new class of insecticides and a material called Sevin onto the market. Honeybee losses caused by the use of this chemical were widespread. The problem was akin to that with lead arsenate. The insecticide killed foraging honeybees slowly, and stored contaminated pollen continued for several weeks to kill the honeybees that ate it. The New York Legislature responded by passing legislation to compensate beekeepers for moving colonies of honeybees away from spray plots being treated for one insect, the gypsy moth. No other state legislatures in the country came to the aid of their beekeepers.

In 1970 the Federal Government responded to the increasing problem by enacting legislation that compensated beekeepers for all pesticide losses. Nearly $40 million had been paid to beekeepers when the program was halted in September 1980. Interestingly, I have not heard a single beekeeper complain about the loss of the program. I believe the reason for this is that it did nothing to lessen the problem. In fact, many beekeepers believed the program was nothing but a facade behind which the perpetrators of problems might hide. There was nothing to prompt anyone to seek a solution.

A new product, Penncap-M, developed by the Pennwalt Corporation, complicated the picture and caused many losses in the late 1970s and early 1980s. Like lead arsenate and Sevin, it kills foragers slowly, allowing them first to collect large quantities of contaminated pollen. However, whereas Sevin might remain toxic in pollen for many weeks, Penncap-M in stored, bee-collected pollen might retain its toxicity for over a year.

At about the same time that Congress enacted legislation compensating beekeepers for losses due to pesticides, it created the Environmental Protection Agency (EPA). No doubt the thinking behind both pieces of legislation was much the same, that contamination of the environment should cease and that government should play a role in this action. Beekeepers, on the other hand, use certain chemicals, including repellents to remove honey, wood preservatives, drugs to treat diseases, and so on. In any event, the American beekeeping community never turned to the EPA for help on the pesticide problem; there is not much doubt that the EPA, industry, the universities, and even many beekeepers looked to compensation as a stopgap and hoped that research would lead to a solution. That research was not forthcoming, for again there was no pressure to conduct it. The EPA did require a better labeling of pesticides than had been in existence earlier, including warning labels indicating that wherever bees were present or endangered certain precautions should be taken. Part of the pesticide problem lies in the fact that honeybees forage over a vast acreage. It is not unusual to have bees from a large apiary collecting pollen and nectar from dozens of farms as much as three miles away at certain times of the year. Honeybee colonies have an efficient system of constantly searching for food that causes them to roam widely. Should they contact pesticide-contaminated flowers, and the beekeeper seek to locate the problem, a long search may be required.

There has been a great deal of research on honeybees and pesticides and certain facts are clear. The only time honeybees are killed in large numbers by pesticides is when they are caught by a spray as they are flying, which is infrequent, or when the flowers they are visiting are contaminated, which is frequent. Covering a beehive with a normal application of an insecticide from a ground sprayer or an airplane has almost no effect on the bees in the hive unless the material is directed into a colony entrance.

Most labels on packaged pesticides that might adversely affect bees state clearly that the material should not be applied when the crop being treated is in flower or when the chemical might contaminate weeds in or near fields adjacent to the area being treated. Enforcement of the precautions stated on the label is in the hands of the states. Some are more aggressive in enforcing regulations than are others; many need reminding by beekeepers of their proper regulatory role.

In 1964 Congress passed legislation that created at each state college a position for a pesticide and toxic chemicals specialist who would serve as a resource person to extension staff, county agents, regulatory agencies, and others in the state. At the time it was recognized that pesticides are important agricultural tools. The person in this position is active in making sure certain pesticides are used properly, as well as in preparing educational materials to advise against misuse and to protect environmental concerns. Some pesticides may be used for special purposes only and can be applied only by certified operators. Beekeepers in areas where pesticides are used should be familiar with the resource people who exist, including the above-mentioned specialist whose job it is to aid them in protecting their bees.

6
JUNE

The major honeyflow in most areas of the northern states will start during late June or early July. The source of the nectar is predominantly clover and the honey is light colored and mild flavored; it is considered a premium honey. Blends of clover honey dominate the table market in the United States and Canada.

However, clover honey is not the only early honey produced in sufficient quantity for a surplus to be harvested in midsummer. Only a few will be mentioned, including raspberry, to a small extent blueberry, and various thistles. Sumac is important in eastern New York, Connecticut, and some other areas. In Pennsylvania, Maryland, West Virginia, and in parts of the west and south, the tulip poplar is an important source of nectar. These trees, which may reach a height of one hundred feet or more, flower in early May near Washington, D.C., so early that most beekeepers will not have built their colonies to sufficient strength to gather a crop. Beekeepers who produce tulip poplar honey like it, though many people find it too dark and strong in flavor. However, one must be careful in criticizing honeys of different flavors; what is pleasant for one is often not for another. I think this has much to do with flavors one learns to distinguish and enjoy in one's youth.

Many factors control what a beekeeper may harvest. Since

bees normally forage over an area two to four miles in diameter, apiaries a few miles apart may produce different quantities of different honeys. Soil type, soil moisture levels, and the weather have the greatest effect on nectar flows. While many people have kept records of varying conditions in an attempt to analyze why plants yield nectar in some years and not in others, no one has any positive, clear-cut idea about what controls nectar production.

DRAWING NEW FOUNDATION

It is a good idea to make a few new combs (frames) each year; in this way one will always have good brood combs that contain predominantly worker cells and also sufficient combs for honey storage. It is impossible to avoid losing a small number of frames each year: some break in the extractor, the ears on top bars break easily, and probably the greatest losses result from damage by mice, wax moths, and other pests.

The best time to draw a new sheet of foundation is during a honeyflow. How frames should be nailed and made is discussed in Chapter 2; here we are concerned only with how and when to put the frames on the colonies.

Beeswax foundation is a delicate material that will stretch, sag, buckle, and warp easily. The best combs are made when the foundation is put into the frames the same day these are put in the hives, but the new frames should be placed on hives only when a honeyflow is in progress and the bees will start to draw the foundation immediately. I emphasize that the honeyflow should be in progress, since it takes a day or two with incoming nectar to stimulate the wax glands of worker bees so that they will secrete wax. Even when colonies have no need to make new comb, some bees will secrete wax during a honeyflow.

The best place to draw foundation is immediately over the brood nest. A favorite method has been to place the queen in the bottom super below an excluder. A super containing four old combs, two on each side, and six new frames with foundation in the middle is placed immediately above the excluder.

Natural comb built on the underside of a cover by a swarm in a box without combs. Note that the combs are evenly spaced but not all perfectly straight. Foundation is used in frames to make straight combs for the beekeeper's convenience; the bees do not care if the combs are straight or not.

Some beekeepers do not use an excluder, since a queen is not inclined to lay eggs in new comb; however, she may do so, especially if the honeyflow is slow, with only a one- or two-pound gain a day. Other supers for honey storage are placed above the super with the foundation.

Wherever one draws foundation it is advisable to use the equivalent of ten frames in a ten-frame super so that combs filled with honey will contain a minimum amount of honey and not break so easily in an extractor. Some beekeepers make new combs by forcing queens to lay eggs in the newly drawn foundation immediately. There is no reason from the point of view of making new combs why this should not be done; however, I think queens will produce more brood in old comb. When one is uncapping new combs filled with honey, it is just as well to cut a little deeper into the comb than one would when uncapping an old comb. This again reduces the weight and lessens the danger of breakage. It is best to take more time to extract new combs, again to reduce breakage. Bees will be able to repair new combs that fracture in an extractor, but it is best to take precautions against this.

I have heard some beekeepers say that new combs of honey should not be used for winter food, but I know of no reason why this should be harmful. I have put up to three new combs filled with honey in the top super (of two) of colonies I was wintering, without adverse effect.

SUPERING

Supers of drawn comb should be added before they are needed. If the honeyflow normally starts on June 15, the hives should be fully supered by that time. One does not add all of the supers at one time, but over a period of four to six weeks. This may mean the colonies will be four to six or more supers high (or the equivalent, if half- or three-quarter-depth supers are used). However, sometimes honeyflows last for many weeks or there is a dribble of honey over a long period of time, in which case more supering may be done, or some supers removed and more added.

When comb honey is produced, one does what is called bottom supering. When first preparing a colony, only one or two section supers are put into place over the brood nest super. Two or three weeks later, if the honeyflow continues, it is time to add a third super. To bottom super, the third super is placed just over the brood nest and under the first honey storage supers that were placed on the colony. The chief reason for this is that the new super will be nearest the brood nest that will contain the young worker bees with developing wax glands and it is here that the foundation will be drawn most rapidly. This super will also be nearest the incoming foragers and will become the chief super for honey storage. There will be less travel stain on the sections above that are being finished.

Some beekeepers think it is advisable to bottom super in the case of extracting honey. I have seen the subject spark a small controversy among commercial honey producers. It is certainly fastest and easiest to top super, which may force the bees to walk through full, or partially filled, supers of honey, but I doubt if this does any real harm; an upper entrance(s) would eliminate the problem of incoming foragers being forced to walk through the brood nest.

[When using ten-frame supers, I prefer to use nine combs in the first one or two honey storage supers I add and after that to use only eight, presuming, of course, that all the combs being added are old ones. It is easy to uncap fat combs and I always seem to be a little short of combs in a good honeyflow. Using eight frames allows me to make up a few extra supers for honey storage.] It may be inconvenient to space the combs when one uses only eight of them. I have never used frame spacers in my supers, though I know many beekeepers who do, because my supers may be brood nest supers one year and honey storage supers the next.

DOES OLD COMB DARKEN HONEY?

[Old comb becomes dark, even black, because it gets stained with pollen and propolis and accumulates cocoons from bees that have developed in its cells. In time, cells may even become smaller because of successive layers of cocoons and the propolis that is deposited in them when they are cleaned. This does no harm and I have many combs in my apiaries that are more than fifty years old.] This horrifies some European beekeepers, who worry that small cells may result in the production of smaller bees (which is true), which will produce smaller crops. There are no data to support this last thought, and if there are some smaller bees in the United States I can only presume that more of them are produced (with a smaller amount of energy being used to produce each bee) and that the honey crop is the same with large or small bees. It is good at this point to repeat what has been said before, that management of the colony is more important than anything else.

A favorite trick of those who advocate using only new

comb for honey production is to place a piece of old comb in a jar of water at the outset of the lecture. After only a few minutes some of the stain can be seen darkening the water and after half an hour the water will be a light brown color. The thought of old comb darkening honey is repugnant and makes a very nice lecture topic. The fact is, however, that while honey is a water solution, ripe honey is not really darkened by old comb. Witness to this fact is that much light honey is produced in this country, most of it with old comb. The physics of what takes place would be interesting to pursue; however, I cannot remember ever having read anything on the subject. High-moisture honey will darken very slightly in old comb, and the extent to which it does has been researched.* Perhaps beekeepers in certain western states who produce only water-white honey would care to pursue the matter, but for most of the country the difference is too small to think about.

MAKING INCREASE

This is the month to think about making new colonies to replace those lost in winter or to increase one's holdings. As discussed earlier, colonies started at this time of year will probably need special attention, especially feeding, in the fall; however, much depends upon the midsummer and fall honey-flow. Small colonies made at this time of year may be used for routine requeening, which is usually done in August (see Chapter 8).

One has two choices in making an increase. A populous colony may be split into four to six smaller units, or a small nucleus colony may be made from stronger colonies showing indications of swarming. Sometimes weakening a strong colony is the easiest way to prevent swarming.

In June, queens from the southern states are relatively cheap, more so than in April and May. To divide a populous colony into several parts one first finds the queen (which is not

---

*G. F. Townsend, "Absorption of colour by honey solutions from brood comb," *Bee World* 55:26–28 (1974).

always an easy task) and places her and a frame of brood in a nucleus box or super on the location of the parent colony. This unit will absorb much of the colony's field force and these bees will protect the brood and the unit should develop into a prosperous colony. The remaining brood, bees, and honey are divided into more or less equal units and each is given a young queen. The queen is introduced to her new colony via the cage in which she was shipped, as described in Chapter 4. It is important that each of the new colonies be given a frame of honey. At this time of year a small colony can usually find sufficient forage for its needs; however, a few pounds of honey should be left in case of inclement weather. The nucleus colonies must be made with a sufficient number of bees to cover and protect brood.

To make a nucleus colony from a strong colony that one fears might swarm, it is again necessary to first find the queen. Either she is picked up and placed in the lowest super or the frame she is on is placed there and the increase is made from bees and brood from the second or third super. When I approach a colony to find a queen, I like to break the colony apart, placing the supers separately on upside-down telescope covers. In this way, when I am searching for the queen in the super I think she is most likely to be in, I am not disturbing the bees in the other supers. Too much smoke or movement of the frames may cause a queen to run off the brood, making her more difficult to find. It is also important to use as little smoke as possible when finding a queen, which means working only in good weather.

I am aware that some beekeepers do not bother to find the queen when making up nucleus colonies. To make the nucleus they open the colony, smoke it, and take what is wanted in the way of brood, bees, and honey, examining each frame for the queen as they proceed. I think there is too much danger of not seeing a queen under these circumstances. If the parent colony is accidentally dequeened, it will probably be lost for honey production that year.

Some of the bees transferred to the new unit will be of foraging age and will probably return to the old colony after collecting food. It is sometimes helpful to stuff the entrance of the new colony with green grass to prevent the bees from flying. The grass should not be packed in too tightly. After twenty-four hours the grass will wilt and a greater number of the residents will reorient to their new home.

Some beekeepers go to the trouble of making up nucleus colonies and then transporting them to another apiary so as not to lose the foragers. Still another technique is to move large colonies that are to be split into several units to another apiary so there will be less drifting after the new colonies are made.

WAX MOTHS

There are several species of insects that may feed on stored comb or on comb in colonies without enough bees to protect it. The most serious of these pests is the greater wax moth. The adult moths have a wingspread of 1 to $1^{1}/_{4}$ inches. Their wings and bodies are a grayish brown that blends well with tree branches. The adult females enter colonies in the late evening to lay their eggs. The larvae that emerge from the eggs tunnel in the comb, where it is difficult for the bees to remove them. Bees in active nests are able to keep wax moths out or at least to reduce the damage they cause to an inconsequential amount.

The greater wax moth, and certain other species that are sometimes a problem, cannot survive much farther north than North Carolina because of the cold weather. In the southern states one must be on guard against them all year; in the north one starts to watch for them in June, though often it is in July or August that they become a problem.

Some wax moths overwinter in the north in comb stored in warm or partially heated buildings. For this reason all comb and comb refuse in the north should be stored where it will be subjected to freezing temperatures. In some areas wax moths are grown as fish bait and adults escaping from colonies where they are grown for this purpose can infest beekeeping equipment. I have seen moths fly from truckloads of bees brought north in the spring.

Wax moths will not infest foundation nor are they a problem with new comb. They can ruin sections of comb honey, not because they consume so much comb and honey, but merely because of their tunnels and piles of frass (fecal matter) on the comb surface. For most beekeepers wax moths are not a problem. Keeping strong colonies is the best defense; the bees in such colonies will easily protect supers of combs on them. Mothballs, especially those made with paradichlorobenzene, can be used to protect combs in storage; combs stored with a fumigant should be aired before being put on colonies.

7
JULY

By the end of the first week in July the intense management of colonies that should have been practiced since mid-April ought to be finished. During this month swarming ceases to be a problem in the north and the bees should harvest and ripen much of the light honey crop. There is still work to be done in the apiary, but it is not as precise as earlier in the season. Daily watching of the scale hive becomes important and fascinating.

USING QUEEN EXCLUDERS

[Honeybees separate their brood and their food, storing the pollen above the brood and the honey above that. If one thinks of the brood nest as a ball spread across four to eight frames, then the pollen is found in a bowl-shaped layer above the brood. The band (bowl) of pollen is a few to ten or more cells thick. This means that under certain circumstances some pollen and new honey may be stored in some of the same frames that contain brood, and the side combs are most likely to be full of honey.]

I prefer to use a queen excluder because it separates brood and food better than the bees themselves. When one puts an excluder on a colony depends upon the swarming season, when the flow will take place, and when one wants to harvest the honey. In my area in central New York State, the optimal time to put the excluder in place is July 7 to July 10, after which my data show that almost no more queen cells will be constructed.

These data are the same anywhere in the United States and Canada north of about the New York–Pennsylvania line. Farther south the swarming season will end earlier and the queen excluders are put in place accordingly. In my immediate area there is usually a short flow from basswood in mid-July. In some years the basswood flow lasts ten to fourteen days, but in one recent year it lasted only three days, with bees storing about ten pounds each day; inclement weather brought the flow to a halt, which was a great disappointment. In that year particularly, had I not put the queen into the bottom super in early July, with the queen excluder above, I would have harvested very little honey.

I have heard beekeepers claim that queen excluders inhibit the movement of bees and hinder ventilation; some call them honey excluders and the subject is often discussed at beekeepers' meetings. However, I am aware that many commercial beekeepers use excluders and would not be without them. While I have heard people complain about excluders, I have never seen any data that indicate they reduce a crop in any way.

I have watched worker bees pass through various types of excluders and have noted they do so with ease. My favorite excluder is the plain sheet of perforated zinc that has been popular so long with commercial beekeepers. It is not carried by some of the supply dealers, probably because of undocumented stories about it. Some claim that the perforations have ragged edges that may cut or tear a bee's wings; I do not believe this. The wire-type excluders were invented with the idea that they would provide better ventilation and ease the passage of the bees. Wood-bound excluders are popular and, while I have no prejudice against them, it seems to me they are more difficult to store and clean.

The easiest way to clean accumulated burr comb from queen excluders is to leave them on top of a hive cover. The sun will melt off the wax, leaving a reasonably clean excluder.

Cleaning a queen excluder by leaving it on top of a hive cover and allowing the sun to melt the burr comb. The melted wax will also protect the cover.

The wax residue left on the cover may help to protect it. Another method is to place the excluders in a solar wax extractor (see page 141 for where to write for plans); still other beekeepers clean excluders with steam.

It is important to store excluders where they will not be trodden upon or damaged. Some queens are smaller than others, and if an excluder is damaged in any way it may be possible for a small queen to move through it. The spacing between the wires or the width of the perforations is critical; it is designed to prevent a queen's moving through because her thorax is wider than that of a worker.

If an excluder is placed on a hive with brood above, it is important to provide an upper entrance so that the drones may escape. Drones have wide thoraxes like queens and cannot move through an excluder. If they cannot escape from the hive, they will die above the excluder, and if they are numerous, they may block it, thus hindering ventilation and the movement of the bees.

UPPER ENTRANCES

During the time when colonies are heavily supered and gathering honey, I often provide an upper entrance. This is done with the belief that additional ventilation is helpful and that foragers using the upper entrance will not congest the brood nest. Where excluders are used, the upper entrance has the added advantage of allowing drones to escape; like certain other practices I undertake in honeybee management, I do not always have proof that honey production is improved, though most beekeepers believe an upper entrance aids ventilation.

There are several ways of making upper entrances. Those beekeepers who drill one-half to one-inch holes in supers for winter ventilation may leave these open for summer use as well. Others push a super back (or forward), providing a long, narrow slit about a quarter to three-eighths of an inch wide. Still others have supers with open knotholes or cracks that allow the free movement of bees in and out. As long as the super is serviceable, these openings appear to cause no trouble. I am writing here about liquid honey production only; beekeepers who produce cut comb or comb honey do not provide ventilation in the storage supers.

SORTING COMBS

There have been two studies that indicate that having large drone populations in a hive does not adversely affect honey production; still, producing a great number of drones does appear to be a waste of energy and thus most beekeepers try to reduce their numbers. July is a good month to remove poor combs, those with an excessive number of drone cells and broken parts, from the brood nest. If an excluder is used, the poor combs with brood can be placed immediately above the excluder; the brood will emerge and the combs may be filled with honey. Later, during extraction, one can determine whether the poor combs should be discarded. Newly drawn combs or old brood combs with only worker cells are used to replace those taken from the brood nest.

If one does not use an excluder, the choice of where to put the poor combs with brood is more difficult. If they are put some distance from the brood nest, in a super one or two supers away from it, the bees may make queen cells, using any eggs or young larvae that are present. When brood is in two separate locations in a hive and only one queen is present, the bees will try to rear a queen in the area the queen does not occupy. A queen will not move from one brood nest area to another when the two are separated by too great a distance. No harm is done when bees attempt to rear a queen under such circumstances, provided that the beekeeper remembers to cut out the cells before they are capped.

VARIATIONS IN HONEYFLOWS

Perhaps it is obvious, but it is worth repeating that there are tremendous variations in nectar or honeyflows. Plants produce nectar solely to attract honeybees and other insects so that pollen may be transferred and pollination take place. Some plants produce pollen in abundance so that a surplus may be harvested by insects; some plants emphasize nectar production, while still others produce both. The weather must be suitable before the honeyflow so that the energy required to synthesize the pollen and nectar is available. Additionally, in the case of nectar, the sugar must be transported through the plant and secreted by the nectary, a gland that usually lies at the base of the flower's sexual parts. Again, this transport of sugar cannot take place in bad weather.

Plants usually secrete nectar over the several days that pollen is being produced and the female parts of the flower are exposed and receptive. In the early spring it must be warm enough for nectar to be secreted and for pollen deposited on the female parts to grow; the same is true in the fall. During summer, hot weather and strong winds may dry nectar so much that it is difficult for bees to collect; rain may dilute nectar. Nectar flows may stop and start again. Although honeyflows have been observed by many people over a long period of time, no one has devised a reliable prediction system; honeyflows are as unpredictable as the weather that controls them.

COMB AND CUT-COMB HONEY PRODUCTION

Producing comb honey is an art as difficult as queen rearing. Much depends on where one lives, for only in areas where there are long, strong nectar flows can one expect to be successful in comb honey production year after year. Some literature states that to make a super of comb honey one merely places the prepared super on a two- or three-story hive. In my experience this may work, but the beekeeper will be successful in producing a super of comb honey less than 10 percent of the time. Long bulletins and several books have been written about comb honey production; the management must be precise and carefully timed.

To produce comb honey one allows a colony to grow in the normal manner in the spring; at the outset of the honeyflow it may occupy three or four supers. One does not prepare the colony or put comb honey supers in place until the first or second day of the honeyflow, since the colony may be prematurely crowded and the bees may chew the new foundation.

On the day the flow starts the queen must be found and placed in the bottom super along with the frames of capped brood. Capped brood is retained in the colony because it will provide more bees for honey production, and when they emerge there is space for the queen to lay. Bees in the egg and larval stage will probably contribute little to the storage of honey and may, in fact, deter it because of the need for bees to feed them. Two comb honey supers are placed on top of the single brood nest super and the rest of the supers are given to other hives.

Under these circumstances the colony will obviously be congested and queen cups and cells may be started. Every frame in the brood nest must be inspected after seven or eight days and the cups and cells removed. The operation must usually be repeated in another week. In the case of a long flow, such as one sometimes gets from alfalfa in late July and August near some of the dairy farms of central New York, even a third inspection may be required. If cells are still found on the third

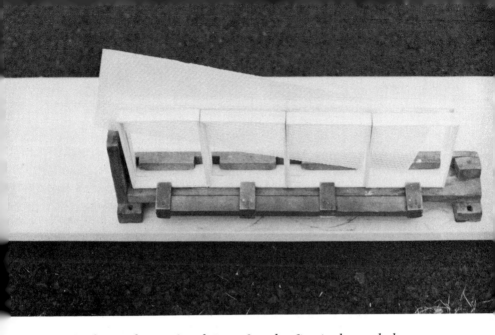

Foundation being fitted into four-by-five-inch comb honey split sections. Making comb honey is an alternative for those who do not own an extractor.

inspection and the honeyflow is still in progress, it is doubtful that the colony will be a good producer.

In unusually good locations one may make more than two supers of comb honey. If a third super is added, it is placed under the first two, but not until one of the first supers is nearly full of honey and the second super at least half full. The new super is kept near the brood nest, where there will be a large number of young bees with well-developed wax glands ready to make new combs.

The storage and marketing of comb honey involve special problems. Wax moths are annoying and even one very small wax worm can spoil a section. Newly made comb honey sections may be put in a freezer for a day to kill any wax moth eggs that might be present. Comb honey must be stored where it is dry, for it will pick up moisture on its surface where the humidity is high. The wax cappings over cells of honey will allow moisture to move in and out; cappings are not moisture-proof. Comb honey is easily damaged by people who "squeeze the product." Most beekeepers who market comb honey place it in plastic bags within window cartons so it is protected by two transparent layers of plastic. Comb honey may granulate rapidly, depending on the type of honey. People who live in areas where there is such honey must plan to sell their comb honey quickly and have it consumed rapidly.

Persons producing cut-comb honey, honey produced on comb honey-type foundation but in half-depth frames, are likely to be more successful than those who produce regular comb honey in square, rectangular, or round sections. The management scheme is much the same; the more crowded the colony the better the production will be. Bee supply manufacturers sell small plastic containers designed to hold a piece of cut comb of about four or five square inches. I would advise novices who do not want to produce liquid honey to start with cut-comb honey.

This is little more than the barest details on comb honey production. A beekeeper can take great satisfaction in producing comb honey, for it is indeed beautiful and delicious. I suggest that anyone who wants to pursue this read one of the many books on the subject, such as Dr. C. C. Miller's *Fifty Years Among the Bees* (see Selected Bibliography). Miller was the most famous of all comb honey producers.

8
AUGUST

In early August I harvest at least some of the light honey. This is the month for requeening and starting to plan for next year. It is also a good month to replace old and worn equipment; often the September and October weather in the north makes working in an apiary difficult and unpleasant. It is too late to think about making new colonies, but the nucleus colonies or swarms hived in June and July will need special attention so they can survive the winter. Swarming is a minor problem in this month. Colonies that swarm in late August and early September are probably trying to rid themselves of surplus bees they might otherwise have difficulty wintering. Ventilation can sometimes be a problem in a hot August, and special attention may be given to making upper entrances to aid the bees both in cooling their hives and in ripening honey.

WHEN IS HONEY RIPE?

Honey that contains more than about 19 percent moisture may ferment. Since nectar may contain from 50 to 75 percent water, the bees use a great deal of energy to eliminate this water. All honey contains yeast cells, which are killed only if the honey is heated to 140°F for thirty minutes, 160°F for one minute, or some combination in between these two values. Honey is hygroscopic, which means it will attract and pick up moisture at its surface if the humidity is high. Honey may have a low moisture content, but if it is exposed to high humidity, it may pick up enough moisture on its surface to ferment, even on the top few thousandths of an inch.

When the cells in the comb are filled with ripe honey they are covered with a thin layer of wax by the bees. The cappings serve to protect the honey from contamination by pollen and propolis and give the bees a surface on which to walk. The cappings over cells of honey appear different both in color and shape from the cappings over cells containing brood. It is perhaps worthwhile to uncap some cells on a frame to be able to learn the difference. When this is done carefully, the bees will easily repair the damage.

It is not advisable to remove honey from a colony during the honeyflow, because it may not be fully ripe; however, if one is dealing with a long flow that continues over a period of several weeks, it may be helpful to do so. The general rule is that one does not remove a frame of honey until it is two-thirds capped. Bees do not usually cap unripe honey; however, the rule is not perfect and depends a great deal on the humidity and how rapidly nectar has been coming into the hive.

In the dryer states of the west and midwest, especially Arizona, one rarely worries about high-moisture honeys. Along the southern coastline, especially in Florida, high-moisture honey and fermentation are a constant worry. In the northeast much depends on the weather. In this area if one removes only frames that are two-thirds capped, and does so only after several days of sunshine, the bees are likely to have reduced the moisture content so that fermentation will not take place.

The moisture content of a honey can be determined only with a refractometer or a hydrometer. Refractometers are available from some bee supply manufacturers; some bee clubs own refractometers that they share among their members. Some state colleges will determine moisture levels for concerned beekeepers who are residents. Nearly all honey buyers will check moisture levels before making a purchase.

METHODS OF REMOVING BEES FROM SUPERS

There is no good or easy way to remove supers full of honey from a hive. Both commercial and hobby beekeepers use a great variety of methods; this fact testifies to the many problems that arise when one "robs the bees." After one goes to the trouble of getting the bees out of the supers, new problems arise. If one removes supers when there is a dearth of nectar in the field, the odor of fresh honey, together with any drops of honey that may fall on the ground, may excite the bees and cause robbing. It is important to keep supers one has removed covered at all times so that they cannot be found by searching bees.

It is also possible that burr and brace comb full of honey will be broken apart when the supers are removed. It is helpful to have some kind of pallet or drip board on which to place the full supers. One may construct a drip board so that it can be moved conveniently with a two-wheel cart.

To remove honey one may use a repellent, a bee escape, or a bee blower; shake and brush the bees off the frames; or wait until cold weather, when the bees may abandon the upper supers. I do not advise this last method, since many honeys will granulate rapidly in the fall. Granulation makes extraction difficult and sometimes impossible; granulated comb honey does not have as fine a flavor as liquid comb honey.

Two bee repellents are now approved for use in removing honey: benzaldehyde and butyric anhydride. Repellent pads are made by constructing a two-inch-high rim the size of a super. The top is covered with two or three layers of burlap or some other absorbent cloth, which is then covered with galvanized iron or a sheet of aluminum painted black to absorb heat from the sun. A small amount of repellent is put on the burlap just before the repellent pad will be used. Whenever one uses a bee repellent, it is best to smoke the bees off the top bars before the repellent board is put into place. Beekeepers refer to this as "starting" the bees in their downward movement.

Butyric anhydride is gaining in popularity as a repellent and I must admit I like the way it drives bees. It is superior to benzaldehyde. Its odor is horrible and this is probably the reason why it is not more widely used. A residual odor is evident in an extracting room for days after honey harvested with it is brought into the building; however, the odor is not absorbed by the honey or the beeswax and does no long-term harm.

Using bee escapes has been my favorite way of removing honey for many years, though they have several disadvantages. Two trips to the apiary are required, as it takes at least twenty-four hours for the bees to leave a super and they will not leave if any brood is present. If one leaves the supers in place above an escape for more than a day, it is important to check them for any holes. These must be plugged or the supers may be

A simple repellent board made with plywood and burlap. The wooden rim, of the same dimensions as a standard super, is two inches deep.

Two men using hive lifters to raise the top three supers so the queen excluder may be removed and a bee escape put into place.

robbed of their honey. When bee escapes are used, and queen excluders removed at the same time, usually three people are needed, two to lift the supers and one to pull the excluder and put the bee escape board in place. If one puts bee escapes in place alone, and breaks several supers apart to do so, then time must be allowed for the bees to clean up any dripping honey from the burr comb broken apart between the supers, which may take an extra day. An advantage in using a bee escape is that one may remove supers early in the morning when the bees are not flying.

Bee blowers, devices that expel a large volume of air through a tube and blow the bees off the comb, are popular with many commercial beekeepers. One should not use a blower unless there is a large volume of honey to remove. Blowers are reasonably fast, though I cannot recall seeing a time–cost study comparing bee blowers and repellents. A few beekeepers use repellent boards and follow up with a bee blower to remove any stragglers. Bee blowers do not make bees as angry as one might think; however, after using a blower for a few minutes there are thousands of confused bees in the air.

Shaking and brushing bees off combs is the oldest method of removing honey; it also makes bees angry and one must be well dressed to use this method. One of its chief disadvantages is that every frame must be handled. A bee brush with nylon bristles may be used and some beekeepers make a brush of grass or plants such as goldenrod. Brushes become stuck up with honey easily, which is a nuisance. If one has only a small number of supers to remove, certainly the shaking and brushing method is the best, despite the furor it creates in the beeyard. A large volume of smoke should always accompany the use of this method.

UNCAPPING AND EXTRACTING

Most beekeepers today produce liquid honey, usually called extracted honey. The term *extracted* is not a good one, for it suggests that something has been taken out of the honey, which is not the case. Rather it refers to the fact that the honey has been extracted, or removed, from the comb.

The first step in extracting honey is to slice off the wax cappings covering the ripe honey cells with a knife. There are special electrically heated knives sold for this purpose, or the old-fashioned method of a long kitchen knife heated in water may be used. The uncapped combs are then placed in a motor-driven machine called an extractor, where they revolve at a high speed and the honey is thrown from the cells by centrifugal force. The honey runs down the inside of the extractor, collects on the bottom, and is drawn off by a spigot.

There are many reasons for producing extracted honey over comb or cut-comb honey. One of the chief ones is that extractors do not damage combs, so the same combs can be replaced on the hives and refilled many times, perhaps for twenty-five to fifty years. Bees must produce much wax to build new comb, and wax production requires honey, energy, and time; it is estimated that ten to fifteen pounds of honey must be consumed by bees for them to produce a pound of beeswax. This honey is "saved" by the beekeeper producing liquid honey. A second reason is that to produce comb or cut-comb honey one must crowd or congest the colony, or else the bees will not fill and cap all the cells and the product will lack a finished appearance. Congestion encourages swarming and comb honey producers must have a more elaborate system of management than is required for the protection of liquid honey. Finally, while most beekeepers agree that comb honey has a superior flavor, it is harder to transport and has a short shelf life. The honey in it may crystallize after three to six months and the coarse crystals detract from the flavor.

Extractors are expensive. It may seem difficult to justify buying a stainless steel machine, even if it has a long life, just to harvest the honey from one or a few hives of bees. Beekeepers report that the cheaper plastic extractors have multiple problems and a relatively short life. Secondhand extractors command a good price. Some beekeepers engage in custom extracting, but others worry that someone using their machine may bring in supers contaminated with American foulbrood spores and that somehow their bees may contact the contaminated honey; I think the chances of this are remote, but it is a difficult argument to combat. Some bee clubs have bought and shared extractors, but sharing equipment is possible only as long as everyone respects the rights of others.

If one intends to stay in beekeeping, even on a small scale, I advise the purchase of a stainless steel extractor, despite the expense. For as long as I can remember, the demand for extractors has always been sufficient that one could recover nearly all of one's investment, even after several years. I would not advise the purchase of an extractor made of anything other than stainless steel.

There are several types of uncapping knives on the market, among which the electrically heated knife is the most popular. Steam knives are used by some beekeepers, but one must have both a knife and a pot-type boiler to generate steam. An advantage of a simple steam knife is that one is less likely (with low- or no-pressure steam) to burn the honey than with an electric knife. Cold knives, knives that are alternately placed in hot water, may be used, but are slower than electric or steam knives. It is possible to uncap combs of honey with an unheated knife, but I think anyone who has done so will soon appreciate the value of a heated knife.

I use only full-depth supers, so that all of my frames are interchangeable. I prefer to use eight or nine combs in a ten-frame super, which means that burr and brace comb build up rapidly. I find it helpful to rotate supers and frames in such a manner that all frames are forced to go through an extractor at least once every three years. This cleans off the extra comb. When I uncap frames, I take the time to remove the extra comb from the top and bottom bars.

I have seen few homemade extractors and I know of no good, readily available plan for making one. I have learned from those who make extractors that a key item is the central bearing that carries the weight; it is the most important part of the extractor. Also, an extractor bearing either should be grease-sealed or it should be designed for use with a special food grease. The diameter of the machine is important; the greater the diameter the slower the machine may run to do an effective job.

HOT ROOMS

A hot room is an insulated room kept between about 90° and 95°F in which one places supers of capped honey a day or two before extraction. Warmed frames of honey uncap and extract faster and with less breakage. If one removes supers of honey from a hive in summer and extracts immediately, there is no need for a hot room, since it is usually warm enough to extract easily. Extracting cold or cool honey is a difficult task and one has to be caught only one time to appreciate the value of a hot room. Some beekeepers keep their extracting room extra warm to keep supers of honey stored there warm; this, however, makes working in the room uncomfortable. The beekeeper with only a few supers to extract should find some way of warming them before extracting starts; the convenience of a hot room or warming cabinet (a miniature hot room) cannot be overstressed.

REDUCING THE MOISTURE IN HIGH-MOISTURE HONEYS

[It is possible to remove about 1 percent moisture from capped honey in twenty-four hours, when the moisture level of the honey is above 17.5 percent moisture, by blowing warm, dry air through the supers.] Removing moisture from high-moisture honeys has been widely practiced for many years; I have seen it done most in honey houses in Ontario, Canada. The beekeepers in that area have been especially interested in making finely granulated honey of high quality, and they need close control of the moisture content for the best quality.

The hot rooms in which water is removed usually have three- to five-foot-diameter fans that force the air down through the stacks of supers. The pallets, or drip boards, must have vents on the sides through which the air can escape. If a large volume of air is not forced through the supers, this method will not work. A small exhaust fan forces moisture-laden air out of the room and a vent allows fresh air in. Neither the exhaust nor the vent should be so large as to make it difficult to hold the room temperature between about 90° and 95°F. Unless the humidity outdoors is unusually high, this temperature will be high enough to dry the air and, subsequently, the honey.

An alternative, when one has only a few supers, is to build a box the size of a super, with a fan in the side to force the air into the box and up through a stack of supers placed over the open top of the box. Some provision must be made to heat the air, and the easiest method is to heat the room or cabinet where the supers are stored.

Once high-moisture honey has been extracted, there is no practical way, on a small scale (less than thousands of pounds), to remove moisture; even when this is done on a large scale, some of the flavor is removed. One of the best ways to circumvent the problem is to blend high- and low-moisture honeys. One can heat high-moisture honey to about 150° to 160°F and bottle it hot, usually with good results. The heating kills the yeast cells that would otherwise cause fermentation. However, once a jar of high-moisture honey is opened, even for a few seconds, airborne yeast cells can enter and the honey may ferment fairly soon.

HARVESTING COMB AND CUT-COMB HONEY

Much comb honey is damaged and even ruined when it is removed from the hive. When bees are smoked, many of them immediately seek honey and engorge; under normal circumstances this is exactly what one would want them to do; however, if the bees are in a super of capped comb or cut-comb honey, they may make dozens of small holes in the cappings to get the honey. If one is removing supers of honey for extracting, this does no harm.

When comb honey is harvested, smoke is not used or only a very small amount is applied. A bee escape works well and many times it can be slipped into place with no smoke or by smoking only the bees in the super below the one being removed. Blowing bees from a comb honey super can sometimes be done without causing them to engorge. It is possible to remove supers of comb honey with a repellent, but the fact that the bees are not started downward with smoke may mean that more bees are near the top of the super and they may become confused. Less repellent is used in this case or the repellent board is put on the super diagonally for the first minute or two.

A last danger in using smoke to remove comb honey is that some black specks of soot may be deposited on the comb surface. In the case of liquid honey these are removed by settling or straining and cause no harm.

REQUEENING

About August 1, one should think about requeening. Whether one should or should not requeen is a question in itself. I know many beekeepers who keep a greater number of colonies, knowing that many will fail during the year because of an old or poor queen, but thinking that it is cheaper, and certainly easier, to proceed in this manner. These beekeepers usually have a number of nucleus colonies—made up of a few thousand bees and a queen in a single super—that they can use at any time during the active season to requeen failing colonies.

Data show that colonies headed by young queens are much less inclined to swarm than those with old queens. Young queens lay more eggs and therefore build bigger populations of bees. For these reasons beekeepers who operate intensively requeen annually. One requeens about August 1, because if the effort fails, there is time to try again.

There are several methods of requeening. None of them is perfect and one must check carefully to make sure the new queen is not killed. If the queen is lost, the bees may raise a queen of their own, which may accomplish the purpose, but this wastes time and may decrease the honey production from a fall honeyflow.

One method of requeening is to find and kill the old queen and to introduce a queen cell or a young queen in a queen cage. I have never seen any data that I trusted as far as this technique is concerned; I suspect it fails about half of the time in the case of populous colonies and that the bees grow a new queen from one of their own larvae. Requeening with a ripe queen cell is better, but I doubt the results are satisfactory more than 75 percent of the time. Placing a ripe queen cell in a colony without removing the queen has been advocated by many beekeepers, but there have been three controlled studies with marked queens that show it is folly; the old queen is not replaced very often.

A popular and usually successful method has been to raise a frame partially filled with brood, with bees to cover, into a

Creating a nucleus colony. This is done by placing into an empty super a frame with some brood, bees to cover the brood, and honey. The new unit is given a queen, which is introduced in the queen cage in which she was shipped. The new colony is placed over the old one, with the hole in the inner cover between the two covered top and bottom with window screening. The new colony has an entrance facing a direction different from that of the old colony (arrow).

The right-hand photo shows the swarm board between the two colonies—a modified inner cover with an entrance cut in the rim (arrow) and two pieces of eight-mesh hardware cloth stapled above and below the hole.

super above a swarm board* or inner cover on the top of the hive in June. The unit is given a new queen (introduced with a queen cage), so that by about August 1 it contains about two pounds of bees and has brood in two to four frames. Such a small unit usually accepts a queen without difficulty. At this time the old queen below is found and killed, the swarm board removed, and a piece of newspaper placed between the two colonies. If the old queen is not found and killed, it is likely that the young one will be killed when the two units are united.

It is tempting to place the nucleus colony on the bottom-board, with the brood and bees from the larger colony above. This is more likely to lead to the death of the young queen. Queen acceptance and rejection are not fully understood, and caution is required whenever one exchanges queens.

THE SECOND DISEASE INSPECTION

The second thorough inspection for American foulbrood is made in early August before the honey is removed. Making an inspection at this time is both difficult and time-consuming because of the lifting of full supers that must be done. However, if one does not use drugs and is concerned with disease control, there is really no recourse. If frames that contain honey but which contained infected brood earlier are removed and get into the extracting routine, they may be scattered in many supers and can cause several colonies to

---

*Swarm boards have been popular for years for both making nucleus colonies and preventing swarming. One may make a swarm board by nailing two pieces of screening (eight-mesh hardware cloth works very well) above and below the hole in an inner cover. Most swarm boards have much larger areas of double screening separated by about one-quarter of an inch. A swarm board allows heat from the stronger colony to move upward, aiding the nucleus colony. The odors of the two colonies are much the same, presumably making acceptance of the new queen easier when the two colonies are united later in the season. Without the quarter-inch separation between the screens, the board will not work. The name *swarm board* comes from the fact that it aids in swarm prevention by weakening the stronger colony.

succumb to the disease. I have known some beekeepers who mark or number their supers and make certain that frames are returned to the same super from which they were removed; some will even go to the extreme of returning the marked supers to the same hive year after year. This requires excessive work, but it is an effective method of preventing the spread of this dreaded disease.

During the American foulbrood inspection the beekeeper looks for other diseases too. As mentioned earlier, European foulbrood, sacbrood, and chalkbrood are diseases that are most likely to develop in colonies under stress; however, it is also true that some races of bees are more susceptible to these diseases than others. If any of these diseases is found seriously affecting a colony in August, when stress should be at a minimum because of high temperatures and plentiful food, the colony should be requeened. [The seriousness of a disease is difficult to define, but any time a beekeeper finds more than twelve to twenty cells of one of these diseases in a single frame, the problem is serious.]

QUEEN EXCLUDERS AGAIN—
THINKING ABOUT WINTER

After the early honey is removed and the colonies have been inspected for disease, it is time to prepare for the fall flow and to think about the storage of ample food for winter. It is generally agreed upon that the best way to winter a colony in the north is in two full-depth supers with a minimum of sixty pounds of honey. It is possible to winter smaller colonies, with less food, and even with sugar syrup; however, doing so successfully requires considerable experience in wintering bees.

For the fall honeyflow I prefer to use an excluder above two supers, putting the excluder in place immediately after removing the light honey, or certainly by the time the fall honeyflow starts, at the latest usually about August 15. In much of the north, where there is a fall honeyflow, goldenrod is the chief source of the nectar, followed by aster. The first goldenrod will start to flower about August 1; however, in my experience, it is usually August 20 when a real honeyflow can be expected. There are several species and varieties of goldenrod that may flower over nearly two months. Most fall honeyflows are fairly long, lasting three or four weeks, but much depends on the weather. If a queen excluder is in place above the second super, the bees will store much of the honey they make during this time just under the excluder, precisely where it is wanted for the winter.

Without an excluder the brood nest is likely to be scattered in the second to the third or even fourth super and honey is likewise found throughout this area. It should be remembered that in many parts of the north, especially the northeastern part of the country, rain and mud can be a problem in September, possibly slowing down yard work. The more the beekeeper can do to speed up operations at this time of the year the better, and in this regard an excluder can be a big help.

9
SEPTEMBER

September is the last month during which thorough colony inspections can be made in the north. The actions taken in August should have prepared most of the colonies for the onslaught of cold weather. I often think of September as the month when one corrects errors made in August; it is not the month to prepare new colonies or to requeen except in an emergency. The northern beekeeper is constantly reminded that thorough inspections of colonies can be made during only five months in states at the latitude of New York and during only six or seven months in states at the latitude of Maryland and Virginia. Those who keep bees farther south have several advantages as far as being able to make closer observations of what is taking place within their hives.

THE THIRD AND FINAL DISEASE INSPECTION

Sometime in mid- or late September the final inspection for American foulbrood is made. As in the case of the second disease inspection in early August, it should be made before the fall honey is removed, and for the same reasons. However, the fall disease inspection should also be regarded as the final inspection for other diseases and of the queen's brood pattern.

Honeybees have the least amount of brood in October and November. Brood rearing may slow down considerably during the latter part of September, making further inspections for the quality of the queen impossible. At the September inspection a beekeeper determines whether or not a colony will survive the winter. It is a waste of honey, and damaging to equipment, to expose colonies that cannot survive the winter to the rigors of that season. Beekeepers who consistently have low winter losses can often credit the high survival rate of their colonies to the fact that the questionable ones were eliminated or combined with others at this fall inspection.

HARVESTING THE FALL HONEY

Fall honey is harvested only after the honeyflow stops. Drying fall honey is often difficult and the bees must be given time. The end of the flow is usually controlled by the date of the first frost, a date usually well advertised by the great number of gardeners trying to save the last few tomatoes and other frost-susceptible vegetables. Aster can sometimes yield a large quantity of nectar after a frost, but it is seldom worth waiting or preparing for the rare year when it does. I have heard of beekeepers harvesting as much as forty pounds of aster honey, but in my experience this happens infrequently.

The methods used for harvesting the fall honey are the same as those used in August, except that on cool evenings bee escapes work better and on cool days it is difficult to use repellents. Brushing and shaking cold bees can precipitate an unusual number of stings.

When the fall honey is removed, it is good to briefly check colonies to make certain they have sufficient food. As indicated in Chapter 10, I like to weigh colonies to ensure there is enough food for winter; however, as one removes supers of honey and the excluders, a quick look into a second super will indicate whether or not the colony has adequate food. If empty frames are seen on one or both sides of a brood nest super, it may be expedient to replace them with one or more full frames of honey. Sometimes when I have new frames of foundation filled with honey, I put them in as side frames. New combs are likely to break in an extractor and if they were spaced ten to the super, they will be narrow enough to fit in as side combs easily. As I stated earlier, bees will winter on new combs as easily as on old ones.

I also save a few frames of honey for early spring feeding. Again it is best to select thin combs that can be placed into a brood nest easily in the spring. Any combs of honey saved through the winter should be stored where they will be subjected to cold temperatures, not inside where it is warm and they may be ravaged by wax moths or where the honey may pick up moisture and ferment. It is also important to protect such combs against mice.

Extracting and processing fall honey will not be covered here, since the problem has already been discussed. Suffice it to say that fall complicates the extracting process both because one is dealing with cold honey that must be warmed to extract and because fall honeys tend to granulate more rapidly than spring and summer honeys. The need for a hot room or a way of warming honey before extraction is more apparent in the fall than in the summer. Fall extraction should not be delayed any longer than is necessary.

UNITING WEAK COLONIES

This section appears here not so much to explain how one should unite weak colonies as to emphasize that uniting can be done this late in the year. I do not mean to say that a beekeeper should put off uniting weak colonies until September, but merely that it is better to do it in September than not at all. It has already been emphasized that colonies of bees, like other animals, can have problems at any time of the year. A small colony with less than two pounds of bees cannot survive the winter; however, two such units, with sufficient food, may.

Uniting in the fall is done in the same way as at any other time (see Chapter 4). The stronger of the two units is placed on the bottomboard and a single sheet of newspaper, with two or three slits six to eight inches long cut with a hive tool, is placed

Uniting two weak colonies. This may be done by placing a piece of newspaper, with a couple of slits cut in it, between the two colonies. The piece of paper that is visible reminds the beekeeper to check the colony the next time he visits the apiary.

over the top boxes of this super. The second colony is placed on top. By the time the newspaper is chewed away, the bees will have more or less the same odor and there will be little or no fighting. One queen will survive. I am not really certain how the second one is killed—perhaps by the bees or perhaps by the first queen. Beekeepers are inclined to believe that the queen from the stronger of the two units will survive, and I cannot argue with this, though I am not certain this is true. When I unite colonies in this manner in the fall, I do not worry about which queen will survive, unless, of course, I know that one is much older, in which case I make an effort to find and kill her.

The bees in the combined units will rearrange their stores for winter; if the bottom super has the greater quantity of honey, this is no problem. The new unit should be weighed in the normal manner in October and fed if necessary.

Uniting two small colonies this late in the fall does not guarantee the bees will survive. The bit of newspaper that hangs over the edge between the supers marks the hive and it should be one of the first checked in the spring.

WASPS AND BEES

Vespula germanica, a new species of yellow jacket that is unusually aggressive in robbing bees, was introduced into the northeastern United States from Europe, probably in the 1950s. Most of the several yellow jacket species in this country will try to enter beehives to steal food, but *V. germanica* is more persistent and successful than the rest. It is especially noticeable in observation hives in the fall; the wasps will take honey or adult bees and may feed on bee larvae, though I have never seen them do so. Most yellow jackets are content to take dead bees from in front of beehives. Honeybees kill some of these invading wasps, but I am certain they do not get all of them.

There is no defense against these wasps. They do not pose a great threat, but they are a nuisance and a curiosity. Finding and destroying their nests is nearly impossible because they are so well hidden. I am increasingly asked questions about them in recent years, as their numbers have appeared to grow. *V. germanica* is the most common species around fast-food restaurants, garbage cans, and picnic benches in August and September. They are not aggressive toward people; however, they do like the same food we do. I have known of several people who have been stung in and around the mouth when they and the wasps were feeding simultaneously on the same food or drink; most of these cases occurred when the person was drinking directly from a bottle or can. There are no repellents that are effective against yellow jackets. Sanitation, that is, eliminating the refuse that may be attractive to these wasps, seems to be the best way to reduce their numbers in an area.

Many people do not know the difference between yellow jackets and honeybees and often blame one for what the other does. A beekeeper has no defense against this, and it is necessary to point out the differences whenever the opportunity arises.

10
OCTOBER

In October, November, and early December bees do little or no brood rearing. On warm October days they search for food and if they find any honey exposed, they are certain to start robbing quickly.

[October is the best month to hunt for bee trees if one enjoys this form of hunting. With luck, one might find a bee tree in one day, though it may take two. A tree is found by catching a bee foraging from late goldenrod or aster in a small box containing a bit of comb and honey (or scented sugar water). If the bee accepts the food, she will return home, tell her hive mates about the newfound treasure, and soon dozens of bees will be flying between the food source and the tree. It is a simple matter to close and move the box a few times until the tree is found. The greatest frustration is to discover that one is tracking the bees to a colony in an apiary rather than to a tree.]

[I mention all this just to emphasize that a good reserve of honey and pollen is very important to the colony at this time. The bees will continue to forage just as long as they are able to do so. It is easy to feed bees in October and they will rapidly take and ripen any syrup that is fed to them. This is the month to give the bees the protection they need for winter.]

WEIGHING COLONIES

Most of my beekeeper friends do not weigh their colonies before winter; they lift the hives from the rear, with one hand in the handhold, and make a judgment on whether or not the colonies need feeding. The fact that they do so will not stop me, however, from talking about and advocating weighing colonies.

A standard bottomboard and cover will weigh four to six pounds each, an inner cover about a pound, and a super with old combs about twenty-two to twenty-five pounds, which makes a total of about sixty pounds. The bees weigh another five to eight pounds, bringing the total weight of an empty hive to about seventy pounds, give or take ten to fifteen pounds, depending on the type of wood used, the age of the combs, the types of covers, and so on. About sixty pounds of honey are required to winter a two-story colony, which means the whole unit should weigh about one hundred thirty pounds. Any colony that weighs less should be fed. Too often when weighing colonies I find some with a total of only one hundred twenty pounds. I do not think anyone can tell the difference between a colony weighing one hundred twenty and one weighing one hundred thirty pounds without a good set of scales. When beekeepers weigh colonies, the weights are recorded on the hive, cover, or inner cover and these figures are used as guidelines to determine how much food to give a colony. I particularly enjoy feeding a gallon of syrup to a colony that weighs only one hundred twenty-five pounds, knowing that this will make a real difference the following spring.

FALL FEEDING

The chief difference between spring and fall feeding is that a richer syrup is usually used in the fall (for methods of feeding see Chapter 4). In the spring, one usually uses one part sugar and one part water, by weight or volume, and two parts sugar and one part water are used in the fall. The difference is not as great as one might suppose: one contains 50 percent

Weighing a hive. Colonies should be weighed in the fall to make certain they have sufficient food; a two-story colony such as the one here should weigh at least one hundred thirty pounds.

sugar and the other about 67 percent. It is a little more difficult to force the sugar into solution in the richer mixture and heating the water is necessary. I prefer to feed warm syrup, either in spring or fall, since I believe the bees start to take the food faster.

There is one method of fall feeding that is different and fascinating to watch. If one places uncapped combs of honey, or combs on which the cappings have been broken, under the brood nest, the bees will remove that honey rapidly and store it where it is wanted. It is best to space four or five combs equally in a ten-frame super for the most rapid uptake of food fed in this way. It is also possible to place syrup in a pan under a colony in the fall; again the bees will empty it rapidly. Bees will not take food in any quantity when it is fed in this way in the spring, and I do not know why this is so.

PREPARATION FOR WINTER

In winter, beekeepers' concerns in the north are to give the bees sufficient food, protection, and ventilation. In the far south the major concern in winter is wax moths and other predators, though even in Florida a colony should have some food reserves. Near North Carolina, the most important concern is giving the bees food and slightly reducing the size of the entrances. In latitudes from Maryland to Pennsylvania, colonies will consume thirty to forty pounds of honey. Mice are a problem in both the north and the south, and precautions must be taken against them.

In areas as far north as New York (exclusive of Long Island) I firmly believe in the use of winter packing. Many years ago beekeepers in my area used large wooden boxes in which they placed two to four colonies with sawdust or some other insulation, but now it is recognized that this much insulation is not desirable or helpful. Bees in my area cluster, generate heat, and easily protect themselves against the coldest weather. However, I am fascinated by the use of even more insulation and heavy sugar feeding by Dr. Don Peer and some of his associates in the Peace River district of the Canadian prairie provinces. Peer has shown that it is possible to overwin-

An interesting and simple way to feed bees in the fall is to place four or five spaced combs of honey in a super under the brood nest. The cappings must be broken for the bees to move the honey. The word feed *was written with a hive tool on the cappings here.*

ter colonies there, where it was once thought impossible. Whether or not it will continue to be economical depends on the relative prices of sugar and packing versus the cost and trucking of the packages from the United States. However, the situation that far north is quite different from that in the northeastern United States and adjacent Canada.

There are two matters of concern. First, the colonies must be prepared so that the bees can ventilate as needed, especially to remove excess moisture. When bees eat honey, they give off large quantities of metabolic water. This will condense on the underside of the cover and eventually cause the whole inside of the colony to become damp and moldy if the bees cannot ventilate properly.

The second problem is that bees that eat honey accumulate fecal matter. Bees will not void fecal matter within the hive unless they have no choice, in which case social order breaks down quickly and the colony perishes. The reason sugar syrup is the preferred winter food for bees is that much less fecal matter is accumulated because it is such a pure source of

Colonies in a poor place for winter. One can see the snow has blown against the supers and into the entrance. Colonies should be protected from strong winds that can cause snow to pile up.

The sequence of steps in packing colonies for winter in black building paper (from Wintering Honey Bees in the Northeast, Cornell Information Bulletin 109, *by E. J. Dyce and R. A. Morse).*

food. Someday beekeepers in my area may find it cheaper and more worthwhile to feed sugar syrup to their bees, as they do in much of Europe, and to harvest the extra honey for sale. At present the price differential is such that it is not worthwhile.

If the top of the colony is wet when it is unpacked in the spring, then the pack did not provide adequate ventilation and the method used should be changed the next year. Some locations can never be improved and are just not good wintering yards; they should be abandoned for winter use.

THE BUILDING-PAPER PACK

For nearly fifty years our favorite method of preparing bees for winter has been to wrap a piece of lightweight black

Heavy binder twine holding the winter pack in place (from Wintering Honey Bees in the Northeast, *Cornell Information Bulletin 109, by E. J. Dyce and R. A. Morse).*

building paper around two colonies;* a second piece over the top replaces the covers. The pack is tied into place with binder twine and laths hold the paper in place around the bottom of the hives. Many beekeepers use slater's felt, but even the most lightweight felt is really too heavy.

This pack is not intended to protect the bees from the cold. The black paper serves to warm colonies on those days in winter when they might not otherwise get a flight; frequent flights in winter are necessary to avoid excessive accumulation

---

**Wintering Honey Bees in the Northeast,* Information Bulletin 109 (9 pages, 1978), by E. J. Dyce and R. A. Morse, can be obtained for a small fee by writing Mailing Room, Building 7, Research Park, Cornell University, Ithaca, N.Y. 14853.

of fecal matter. The black paper also aids in melting snow from immediately around colonies.

A major problem in wintering bees anywhere is to get rid of excess moisture that may accumulate in the hive. I leave the holes in the inner cover open or drill a hole in the upper super to make an upper entrance. A bun of straw (wheat straw is best) is placed on top of the two colonies to form a peak in the center to aid in shedding rain and snow; the regular hive covers are stored indoors for winter. An added virtue of the winter pack is that it protects the wooden equipment.

In the Cornell apiaries a hive stand six inches high is used that will hold two colonies. In the summer the colonies are about six to ten inches apart; in the winter, when they are wrapped, they are pushed together. The hive stand is constructed so that there is a dead air space under the colonies in winter.

The building-paper pack is cheap to make. If the paper is precut and all the materials ready ahead of time, it takes six to eight minutes to wrap and prepare two colonies for winter. The success of the winter pack is determined in the spring (see Chapter 4 for more discussion on this point).

11
NOVEMBER

In some years I have been behind in my work and forced to continue fall feeding into November. This can be done, but it is certainly better to complete the job earlier; in many years November feeding is difficult. Populous colonies will take food at this time of the year far better than small ones; in fact, it is almost impossible to get the small units to consume food late in the fall. If the bees do not remove all the syrup from the feeder jars in a few days, it is better to mark the colonies and try to start spring feeding earlier than normal.

Mice move into active beehives in late October and early November, when there is a heavy frost almost every night, with snow starting between November 15 and 20. I have seen bits of wax, obviously chewed out of combs by mice, at the entrances on the bottomboards at this time of year. It is a bad sign and indicates that I am late in distributing the mouse poison.

In general, November is the time to think about honey shows, selling honey, and reviewing the season's work. It is also the time to place orders for packages and queens for the following spring. The better package bee and queen producers are booked up early.

Typical mouse damage to a comb. Usually mice will chew away an area much larger than this, sometimes six to eight inches in diameter.

HONEY SHOWS

I like honey shows. More important than anything else, they are an opportunity for beekeepers to establish a reputation for quality. Awards from honey shows may be used for promotion, in addition to giving a winner satisfaction. I sometimes feel bad for newcomers when the same people win show after show, but this also indicates that the winners are experts. In my experience those who do well at shows also market a superior product.

The object in a honey show is for the beekeeper to put into a jar or package the same high-quality product that the bee puts into a hive. I, and probably most judges in a honey show, emphasize those factors over which the beekeeper has the greatest control.

[I am especially disturbed when I see a ring of foam on the top inside of a jar of liquid honey, and it is even more disturbing to see specks of wax or other foreign material on the top of the foam. Foam on honey results from air that is added, usually when the honey is extracted. A honey pump may also add air if it is not properly used. Air is removed by settling at a moderate temperature. Liquid honey should be free of lint that may be added when using a straining cloth or which may settle with dust from the air. The moisture content of liquid honey is important, but less so, since it is one of those factors over which the beekeeper has little control; I award an extra point or two for low-moisture honey, but it is not a big factor in judging. Granulation is another factor the beekeeper controls, and honey that shows signs of granulation is discounted heavily and usually disqualified.]

Comb and cut-comb honey are easy to judge because so many things can go wrong in their production. The worst fault is travel stain; this is the yellowing of the comb surface caused by the deposition of bits of propolis and pollen from bees walking over the comb. Unfilled and uncapped cells can be a problem with comb honey. In a perfect section all cells should be capped and the cappings should be uniform. Of course, one rarely has perfectly capped sections and so the judge is left to count the number of uncapped cells and to compare the entries. Many beekeepers make the mistake of using smoke to remove comb honey. Smoked bees may perforate the cappings in an effort to feed on honey, or bits of soot may be left on the comb surface. Unclean wood on comb honey sections can be a serious problem; it is easy for wood to become stained with propolis. Round sections of comb honey are often difficult to judge because they are usually well capped and the plastic in which they are contained is protected against staining by propolis.

Beeswax, like comb honey, is easy to judge. One seeks a yellow wax with a good odor. The cake should be free of cracks and layering should not be visible. As beeswax cools, any dirt that is present settles to the bottom, where it is easily seen.

I encourage people to enter honey shows. Lectures on

EASTERN APICULTURAL SOCIETY
JUDGE'S SCORE CARD

Event: <u>EXTRACTED HONEY</u> Class: _____ Entry No.: _____

Point scoring	Item	Judge's remarks
20	*Density* (above 18.6 disqualified)	
10	*Freedom from crystals*	
30	Cleanliness a. Lint (7) b. Dirt (10) c. Wax (7) d. Foam (6) 30	
30	*Flavor* Absence of, overheating, fermentation	
10	*Container appearance* Cleanliness and neatness	
100	Award:	

how to prepare honey for show are given more frequently today. Two points are important: The entries should be numbered so that the judge cannot identify the person making the entry and one should insist that the judge fill out a score card so that he is accountable for his decisions. The judging card for extracted (liquid) honey used by the Eastern Apicultural Society (EAS) is illustrated. While it may not be perfect, the relative assignment of points is considered good. The EAS cards are being used by others outside the society.

PACKING HONEY

This section is concerned primarily with the preparation of liquid honey for market. The object is to offer a product that will have a shelf life and eye appeal for four to six months. Several firms make a variety of equipment for the preparation of honey for one's own use or market. [I feel that only equipment made of stainless steel is worth purchasing. Many beekeepers use secondhand stainless steel equipment to good advantage.]

[The two chief problems found by the honey packer are fermentation and granulation. Those who use flail-type uncappers may have an additional problem removing beeswax particles from the liquid honey. Heating the honey will destroy both the yeasts and the crystal nuclei on which new crystals might grow. If honey contains bits of wax, one must be careful that it does not melt, as beeswax dissolves easily in honey. At temperatures between 100° and 120°F straining is easy and the wax will not melt. Persons who prefer or pack unheated honey must understand the problems and dangers involved in honey storage. The primary ones are that even slight fermentation will make a honey unpalatable and there is no question in most people's minds that honey with coarse crystals does not have as fine a flavor as liquid honey or one with fine crystals.

The honey on the average grocery shelf today is bulk-heated to 140°F for thirty minutes, flash pasteurized to 160°F for one minute, or subjected to some combination of these times and temperatures. In both cases the honey is packed hot. If the honey is allowed to cool before it is put into jars, the dust from the air that settles into jars will be sufficient to cause premature granulation. Many honey packers today filter their honey, which removes pollen and some small portion of the food value. It does, however, increase the shelf life and does not have an adverse effect on the flavor.

Light-colored honeys can be heated to high temperatures and, because of their low protein content, will not have a burned flavor. However, dark honeys suffer greatly from overheating and often have a burned flavor. It is almost impossible to find a jar of buckwheat honey on the market that has retained the flavor it had when first removed from the hive.]

LABELING HONEY

Labels for honey jars are available from several of the bee supply houses; some are attractive and complement the jar and color of the honey. A person interested in large-scale packing is advised to research this area carefully. People have strong preferences and many studies have demonstrated that the sales of a product are greatly influenced by color combinations and the proper use of words.

I think beekeepers could do a much better job of telling the public about honey and especially about some of the delightful honey flavors that exist. I suggest that special labels that describe the type of honey, to be used as second labels on a jar, would be useful. One of my favorite honeys is basswood. It is light in color but has what I call a strong, minty flavor. Some of my friends who are not familiar with honey have thought there was something wrong with basswood honey when they first tasted it; however, after a brief explanation, some (though, I admit, not all) of them commented that they did enjoy the flavor. Whenever I have had honey tastings, I have been able to please everyone by offering a variety, and it seems that the more information people have, the more likely they are to find a honey they enjoy.

[Neither federal nor state agencies concerned with public health and the food industry pay much attention to the beekeeping industry. Honey is as safe and pure as any agricultural product on the market. No microorganism can grow in honey.] Agencies concerned with these matters have put their attention elsewhere, which also means it is up to the individual beekeeper to ensure the high quality of his product.

RENDERING CAPPINGS AND OLD COMBS

Melting down beeswax and beeswax refuse is a messy, dirty, and dangerous job. Many beekeepers refuse to render their own beeswax for these reasons. They package it and send it off to one of the companies that are glad to do it for a fee. Several bee supply companies will pay a premium for beeswax refuse when it is exchanged for equipment.

Hot water, steam, or a solar wax extractor may be used to render cappings and old comb. Beeswax melts at about 148°F, so high temperatures are required. One should never heat a container of wax or wax refuse over an open flame. Beeswax is easily ignited and there have been many disastrous fires as a result of attempts to render beeswax or make candles over an open flame.

Solar wax extractors do a good (though not perfect) job of rendering cappings and a fair job of rendering old comb. (The cocoons in old comb seem to retain the wax.) A beekeeper who uses a solar wax extractor should save the refuse, which can be easily molded and conveniently packaged while it is still hot. Plans for making a solar wax extractor are available at no charge.*

The only way to separate beeswax effectively from the debris is with pressure applied to a cheese (the name given the wax refuse wrapped in burlap or some other heavy cloth used for pressing) under water; however, pressure is of no value without time. Beeswax does not separate easily from the old comb and its contaminants and most wax presses are allowed to run for at least ten hours, with additional pressure applied every hour or half hour.

Wax presses are not available from bee supply companies and all those in use are homemade. I know of no plans for making a wax press and suggest that anyone wishing to build a unit work with a beekeeper who owns one. One may find mention of wax presses in some of the older bee books. Hot, liquid beeswax in contact with iron, brass, zinc, or copper will darken; iron is the worst offender, followed by the others in the order listed. As in the case of honey, it is best to render beeswax using stainless steel.

---

*Solar beeswax extractor, Agriculture Extension Service, Pennsylvania State University, University Park, Pa. 16802.

12
DECEMBER

In most places in the United States there is little activity in the beeyard in December. Even in the far south it is usually too cold to rear queens, and while some flowers may be in bloom, there is little bee flight. In the north it is too late to feed bees or to make up for work not done earlier. During this month bees begin to rear brood, though the peak in brood rearing is months away. This is a good month to repair equipment, trim trees and bushes around apiaries, and make plans for spring.

I will use this chapter to review some bee biology and some aspects of beekeeping that did not fit conveniently into the preceding chapters.

HONEYBEE LIFE HISTORY

There are four stages in the development and life of a honeybee: egg, larva, pupa, and adult. Honeybees are different from all other insects in that they control their brood nest temperature within close limits. As a result, the young develop in precise periods of time. It is being discovered that races of honeybees vary greatly and that some races, notably those from Africa, have a shorter development time; some may develop in one or more days less than the average European worker. The table shown here has been a standard for many years for European honeybees.

Honeybee development time (in days)

Stage	Worker	Queen	Drone
Egg	3	3	3
Larva	6	$5^1/_2$	$6^1/_2$
Pupa	12	$7^1/_2$	$14^1/_2$
Total	21	16	24

Abnormalities sometimes occur in honeybees, as they do in all animals. Brood may be slightly chilled, not enough to cause death but sufficiently to cause the malformation of certain body parts such as the legs or wings; such bees are usually removed by the rest of the worker force. Malnutrition may cause some bees to be slightly smaller than others, especially queen bees. Queens vary far more than workers or drones and may sometimes have some almost workerlike characteristics. Beekeepers who grow queens take special precautions to make certain the developing larvae are especially well fed.

An egg laid by a queen may develop into a worker, a drone, or another queen, depending upon the circumstances. If the egg is not fertilized, it develops into a male. The queen controls whether or not sperm are released from the spermatheca and may fertilize the egg. The research that has been done indicates that she measures the size of the cell in which she is about to lay an egg. If the cell is the larger drone cell, no sperm are released. Whether a queen or a worker develops from a fertilized egg depends upon the type of cell in which it is deposited (worker cell or queen cup) and the food fed to the larva. Worker and queen larvae receive the same food for the first two days of larval life, but afterward the food fed the worker larva is less rich and contains more honey. This change in diet causes the many visible differences in form between the worker and the queen. Honeybees that have lost their queen can rear a new one using a one- or two-day-old worker larva.

Mating in honeybees takes place three to five days after the queen emerges. Under normal circumstances, queens take two or three mating flights and mate with six to eight drones over a period of one to two days. Mating in honeybees is still not well understood and some recent research suggests that queens may mate more frequently. Weather especially may delay mating. During the few mating flights she makes, the queen receives and packs into her spermatheca, a saclike organ, several million sperm, which will last her the rest of her life. Queens never mate again after they start egg laying. Drones are not mature and capable of mating until they are about twelve days old. The males die in the act of mating and fall to the ground. Worker honeybees are not capable of mating, this being one of the many ways their bodies are different from those of queens.

The length of time bees live varies greatly. Queens may live two to five or more years; drones probably live six to eight weeks, though in the fall the workers stop feeding them and they are slowly driven from the hive. Drones do no work and are of no value in cool or cold weather; they are not found in a normal colony in winter. In fact, seeing drones in a colony in the late fall, winter, or early spring indicates that something is wrong with the colony. Worker bees live five to seven weeks in the summer when they are foraging, and longer in the winter, four to six months, when they are not working so hard. Worker bees die from hard work, though they may also suffer from several diseases and physical damage to their bodies from the environment. Older worker bees will usually have some part of their wings frayed.

If a colony loses its queen and for some reason cannot rear a new one, the small ovaries in a few workers may develop and these workers, called laying workers, produce small eggs. These eggs may fail to develop, but if they do develop, the progeny are more or less normal males capable of mating. Only under very rare circumstances will an unfertilized worker egg develop into a queen.

IDENTIFICATION AND CONTROL OF PESTS, PREDATORS, AND DISEASES

American and European Foulbrood

One is always suspicious of larvae that are anything other than a glistening, pearly white color. Likewise, one always examines the contents of any capped cell where the capping is sunken, discolored, or perforated, for this indicates there is a dead pupa within the cell and a disease.*

American and European foulbroods (AFB and EFB) are diseases that are often confused. It is not uncommon to have a mixed infection, making diagnosis difficult. When one finds only one, two, or three cells of foulbrood the problem is often still more difficult, for there is little material to examine.

In general, EFB kills the developing bee in the larval stage, before pupation begins. Thus the larvae killed by these bacteria are usually curled in the bottoms of their cells; they may be off white, brown, or black, depending upon the extent of the decay. Larvae that die from American foulbrood are usually not killed until pupation has begun or is even well advanced; the larva lies flat on its back in the bottom of the cell. Often pupation is far enough advanced that a leg or the tongue is developed; these appendages may stick up from the bottom of the cell. While brood dead from the two diseases may overlap in age, one does not see a leg or tongue sticking up in the case of EFB. Larvae dead from AFB may be of almost any color, as in the case of EFB. While larvae dead from AFB are still drying, the remnants of the dead animal are a sticky, gooey mass that will stretch and snap when punctured with the corner of a hive tool or a toothpick. Some people say the two diseases have distinctive odors, but I have never thought this was a reliable method of diagnosis.

One may send a piece of brood comb, about four inches square, to the Bioenvironmental Bee Laboratory, U.S. Depart-

---

*There are several good references that one may use to identify bee diseases. See the Diseases section of the Selected Bibliography for details.

ment of Agriculture, Beltsville, Maryland 20705, for free diagnosis. This service has been available for many years in the continuing effort to control American foulbrood. Samples should be wrapped in paper, not foil, plastic, or waxed paper, which would retain moisture and cause mold growth. The wrapped samples should be sent in a wooden or heavy cardboard box clearly marked with a return address. The laboratory will accept samples from anywhere in the world.

When my students and I find a case of AFB, which happens every so often, we dig a two- or three-foot-deep hole in which we burn and then bury the combs, inner cover, and bees. The bees are first killed, and the supers, bottomboard, and cover scorched. Other bees in our area receive a routine state inspection each year; in addition, we inspect some of our neighbors' bees, especially near apiaries where critical experiments are under way. The result is that the disease incidence is low in our area and burning is the cheapest, easiest method of disposing of it. Also, of course, we do not want any drugs interfering with our experimental work. One must determine what the state policy is and, more importantly, how much disease is in the area. In some states, where valleys and mountains offer isolation, it is possible to rid an area of disease, much to everyone's benefit. Such an effort may require cooperation from neighbors, which is not always easy to obtain. The last resort, of course, is to use a drug to control American foulbrood. Directions accompany these materials, which are sold by nearly all bee supply dealers.

Other Brood Diseases

Two other diseases that may affect bees in the larval and early pupal stages are sacbrood, caused by a virus, and chalkbrood, caused by a fungus. Chalkbrood was first found in the late 1960s in this country, so it is not mentioned in the older American bee literature. Neither of these diseases affects the egg or the adult.

Sacbrood gets its name from the fact that a larva killed by

the virus may be lifted from the cell easily and has a saclike appearance. When the sac is broken (and it breaks easily), the contents are watery. Chalkbrood gets its name from the fact that the dead mummies are usually covered with white, fuzzy fungus. When the fungus forms spores, the white mummies turn gray or black.

There is no chemical control for either disease. Both diseases, like EFB, are more common in the spring. All three are often found in colonies that have been stressed by a pesticide loss; they are called stress diseases and it is believed that their control, except under the stress of pesticide loss, is possible through selecting good locations. All three of these diseases are much more common in apiaries that are damp, poorly drained, and shaded.

Adult Bee Diseases

Nosema is the only disease of adult bees that is of major concern in North America. It is caused by a microscopic, one-celled organism that invades and destroys the cells in the gut of adult honeybees. It is difficult to diagnose since the symptoms are the same as with old age, pesticide poisoning, or overheating. If one pulls the large intestine from the abdomen, it will be swollen and white. A normal intestine is brownish and one may see the concentric rings, which are not visible in an infected bee.

Nosema is also a stress disease and the same precautions that will prevent larval stress disease will control nosema. One drug is available and is widely sold for the control of nosema, but I have never recommended its use. The reason for this is that while one may greatly lower the incidence of nosema through drug use, this does nothing to relieve the stress that caused the disease and may encourage the development of larval stress diseases. I think if I had bees in some of the northern parts of Canada I might use the drug at certain times of the year. However, this is a question that is better answered by the commercial beekeepers in one's area.

Chilled Brood

[On an especially cold night in the early spring bees may be unable to protect all of their brood and some may be chilled and die. Chilled brood is also often evident after a pesticide loss. When eggs and larvae are chilled and killed, they are removed rather promptly by house-cleaning bees; however, it may take many days, even weeks, for bees to uncap and remove dead pupae. Chilled brood may be identified by the fact that it is found on the edge of the brood nest. One may find only a few cells of chilled brood or there may be large patches involved.]

One should note that no disease of honeybees has an effect on any other animal. Diseases of American honeybees, which are European in origin, may affect Asian honeybee species and vice versa.

MICE, SKUNKS, AND BEARS

Mice, skunks, and bears cause great damage to bees in many parts of North America. Mice in active beehives, especially in the fall and winter, are probably the most destructive animal with which the beekeeper must contend. Mice, rats, and sometimes squirrels may nest in stored supers in outbuildings. The best way to protect against such damage is to reinforce buildings so that they cannot be entered.

Mice are able to walk into a beehive during cool fall weather (when the bees are clustered), chew away comb, and carry in grass and refuse with which to make a nest. Apparently whatever vibrations and/or noise are made are not sufficient to cause the bees to break their cluster and attack the invaders. I have seen mice that have spent the winter in comfort inside a warm, dry beehive and successfully reared a litter of young before their nests were torn apart and they were driven from the hives when the bee populations grew in May and June.

[Mice apparently do not eat honey, pollen, or bees; their concern is a place to nest, and they seek food outside the hive. It is the construction of their nest that causes the damage. Nests vary in size, but they are rarely less than six inches in diameter and may be larger. To make a nest the mice remove and destroy comb from several frames. Too often when the bees rebuild this comb, they make drone comb.

Skunks usually feed on bees at colony entrances late in the evening. They walk up to an entrance, scratch on it, and when bees come out to investigate, the skunks swat them, usually stunning them, and eat them. Skunks prefer to feed on colonies with small populations, since such hives have fewer guards and the skunk is not stung excessively. Those who have watched skunks feeding in apiaries report that they move up and down rows of hives until they find one that would appear to have only one guard at a time at the entrance. The grass in front of hives where skunks feed is torn up and mud is usually visible on the hives. This is caused by skunks scratching themselves and their surroundings when they are stung. Skunks feeding on bees have been dissected and stings have been found in their mouths, esophagi, and stomachs; these stings do not deter the skunks from feeding. Observers report that mother skunks may teach their young to feed at beehives.

Most beekeepers with skunk problems use poison to control them, but effective poisons are less readily available than they were a few years ago. An electric fence, with one or two wires close to the ground, is reportedly an effective way to deter skunks. Placing beehives on hive stands has been tried with some success, but high hive stands are a nuisance. One may trap a skunk, but most trapped skunks will release their familiar odor, making it nearly impossible to work in the apiary for weeks. If their ground holes can be located, trapping skunks away from the apiary is reportedly effective.

Bears are destructive in certain parts of the country. Bears like honey and brood, though there is some argument as to which they prefer. Bears may destroy one or many hives when they visit, though usually once a bear starts to feed in an apiary, it does not stop until all of the hives are destroyed. To obtain food they knock over hives, dump frames on the ground, and sometimes carry hives and frames several hundred feet from an apiary. As in the case of skunks, the fact that they are stung does not prevent repeated feeding.

Game departments and state legislatures vary in their

attitudes toward bears. In some states, bear populations are controlled so that they are not a problem. Some states and Canadian provinces compensate beekeepers for bear damage, but most refuse to admit their bears are a problem. This has led to some unpleasant confrontations and desperate actions on the part of beekeepers who have suffered losses.

Some states allow the shooting and trapping of destructive bears. One northeastern state allows bears to be trapped, but

Bear damage. When a bear attacks an apiary, it may destroy one or all of the hives. Frames and hive parts will be found scattered, sometimes a hundred or more yards from the hive.

does not state what should be done once the bear is in the trap! Electric fences are used effectively in some areas to deter bears; however, beekeepers agree that bears accustomed to feeding on bees will not let an electric fence stop an attack.

HONEY PRODUCTS

About half of the honey produced in the United States is used as table honey and the remaining half is used in the commercial baking trade to make certain graham crackers, cereals, cakes, and cookies. Only about 1 percent of U.S. honey is used in specialty products, though there are some delightful ones on the market.

Honey butter is perhaps the best known honey product; it is a combination of honey and butter. The commercial process is a secret one developed in the late 1930s. Ordinarily when one makes honey butter the product will turn rancid in a couple of weeks. The commercial product, which has never been successfully duplicated, is sold from the refrigerator case in grocery stores, but has a long shelf life. One may make honey butter at home by mixing crystallized honey and butter. The proportions may vary, but a mixture with 10 percent butter is satisfactory; how much butter one uses is a matter of taste. If the mixture is kept in the refrigerator, it will last two to three weeks. As in the case of crystallized honey, the finer the granules the better the flavor of the honey butter will be.

Honey jelly is seen from time to time in honey shows. At one time it was made commercially, but the product never gained popularity. Most of the recipes I have seen for making honey jelly are not good. The rigidity of the jelly depends upon three factors: the concentration of the sugar, the concentration of the pectin used, and the acidity of the honey–water mixture, which is expressed in terms of its pH. The optimum sugar concentration, or in the case of honey the total solids content, should be 67 percent; the most satisfactory pH is 3.0 and at a lower pH the jelly is not firm. In the manufacture of commercial lots it would probably be advisable to use a pH meter. Acid, usually tartaric, is added to lower the pH and make the honey more acid. In addition to correcting the pH, this acid also makes the honey jelly more tart and many persons believe it improves the flavor. It is possible to use lemon juice to lower the pH; this is a satisfactory method of obtaining the correct degree of acidity and at the same time gives an excellent flavor. Limes would have the same effect. The use of juice of this sort should have a commercial appeal as well as being satisfactory for making jelly at home.

Honey wine (mead)* is probably the oldest alcoholic drink known to man. To make mead, one dilutes honey with clean spring, well, or rain water, adds nutrients and yeasts, and allows the mixture to ferment. To make a sweet mead one uses about four pounds of honey and one gallon of water; a dry mead is made by using only three or three and a half pounds of honey. When one dilutes the honey, the nutrients in it are also diluted and must be replaced by adding four grams of ammonium phosphate and four grams of tartaric or citric acid for each gallon of water (a gram is one twenty-eighth of an ounce). One may add one quart of apple juice for each gallon of water instead of the nutrients; apple–honey meads have been famous for centuries. Many mead makers boil the honey–water mixture before adding the nutrients (or juice); this prevents the mead from becoming cloudy when it ages. Any yeast that will make a good sauterne-type wine will make a good mead. A good mead should age at least a year and those aged several years are better. It is also possible to make a sparkling, champagne-like mead at home.

Honey vinegar has never been widely made, mostly, I suspect, because apple cider vinegar is so cheap and easy to make. To make vinegar one first makes wine; however, much less alcohol is needed and only one and a half pounds of honey per gallon of water should be used. When the alcohol fermentation is finished, one adds mother of vinegar either from a culture or by just leaving the bung in the barrel or hole in the fermentation jug open and depending upon airborne bacteria to start the second acetic acid fermentation. When home vinegar making was popular, people used the same barrels year after year and could depend upon the bacteria that convert the alcohol into vinegar being present all the time. The alcoholic fermentation usually takes two to three weeks, but the conversion of alcohol into acetic acid is a much slower process and may take several months. As in the case of honey wine, it is

---

Making Mead (128 pages, 1980), by the author, is available from Wicwas Press, 425 Hanshaw Road, Ithaca, N.Y. 14850.

best first to boil the honey–water mixture; otherwise the final product may be cloudy. Nutrients should be used, as they were in making mead, and are added after the honey–water has been boiled and cooled.

Honey has been used in a wide variety of other products ranging from drugs and cough medicines to beauty products. It is mentioned in a great number of ancient Egyptian remedies, probably because its sweetness covered up some of the harsh taste found in certain plant extracts. Honey has been used in curing tobacco and in packing and flavoring meat products. Despite the great number of recipes that have been developed, the total quantity used in other products, outside of baked goods, remains small.

NATURAL HONEY

So-called natural food is difficult to discuss and define. There are no legal, official, or even unofficial guidelines as to what is meant by such words as *organic, raw, uncooked,* and so on, on honey and other food labels. Government agencies that supervise food quality have steadfastly refused to get involved in this issue, one reason being that the proponents of natural foods have been unable to agree among themselves on definitions. By the dictionary's definition, the word *organic* means "of plant or animal origin," so it can be applied to any food. Some who call themselves organic farmers are willing to use natural insecticides, such as rotenone and pyrethrins; others are not. The development of synthetic pyrethrins complicates the question even more. Are growing practices using manmade pyrethrins any less organic than those using natural ones?

Certainly all honey can be called organic; however, there is the question of the temperature to which one may heat it and still call it raw or uncooked. Is the maximum temperature 90°, 95°, 100°, or, as some maintain, 120°F or higher? I have even heard some beekeepers say that to them raw means unfiltered and that it has nothing to do with the temperature to which the honey is heated. There is no question that the flash-heating of honey does much less damage than slow heating over a long period of time—thus one cannot just think in terms of a maximum temperature. When one sees a jar of liquid honey on the grocery shelf, even if it is labeled "raw," it has probably been heated to a fairly high temperature, for otherwise it would not still be liquid.

It is much easier to understand the terms *extracted, strained, filtered,* and *pure* as applied to honey. The machine used to remove honey from the comb is called an extractor. I do not know who coined the word, but it is a poor one, for many people think the term *extracted honey* refers to honey from which something has been removed. Of course, this is not true. There have been feeble attempts to rename the honey extractor, but all have failed. Beekeepers generally talk about *extracted honey* among themselves, but do not use the term on labels, preferring to say *liquid honey.*

Most beekeepers strain their honey by passing it through cheesecloth or nylon. This removes the coarse particles of wax and comb. A small amount of lint is added when cheesecloth is used and for this reason nylon is preferred. Filtration, as it is generally understood, involves pumping honey under pressure through a filter press using diatomaceous earth (earth composed of the shells of tiny sea creatures called diatoms). Under a microscope they look like miniature sea shells such as one would pick up from a beach; they are almost pure silica. When honey or any food product is pushed through a layer of diatomaceous earth, all of the small particles in it are trapped by the diatoms and removed; there is no adverse effect on the flavor and the result is a crystal clear product. Examples of filtered foods include vinegar, nearly all alcoholic beverages, many fruit juices, though obviously not those containing pulp, and almost any other liquid that is clear. Filtered honey remains liquid because it has been heated and the particles on which crystals can develop have been eliminated; this results in a longer shelf life. Filtration also removes all pollen particles, which may be unfortunate, but if it is done properly, the flavor of the honey is not damaged.

The use of the term *pure* by beekeepers dates to about 1880 or 1890, and perhaps earlier. At that time sugar production was much lower than it is today and honey commanded a much higher price. Before the "pure food and drug" laws were passed by Congress in 1906, food adulteration was neither illegal nor uncommon. Beekeepers resorted to the use of the word *pure* to indicate that their product was different and not adulterated; it is a term that has stuck with the industry ever since. In some ways it is misleading to use the term today, for it might give the impression that there is honey on the market that is not pure. Today we do not need the term and I think it should be dropped.

BEESWAX PRODUCTS

Beeswax is a remarkable product. It is eaten with comb honey; it is used in paint, skin creams, and lipsticks, and as a water repellent, polish, and rust inhibitor. The Egyptians used beeswax in their embalming process and it is also used to wax thread. The greatest user of beeswax is the beekeeping industry, which needs it to make comb foundation; the second greatest user is the candle industry. Beeswax has many virtues in candle making. It gives a pleasing, fragrant odor when burned, unlike the strong, sometimes rancid, odor evident with certain animal fats. Beeswax candles produce almost no smoke, unlike those made from paraffin. Beeswax melts at a higher temperature than many waxes and thus candles made from it hold their form better. Many religious groups insist on pure, or almost pure, beeswax candles in certain religious events, saying that beeswax, being the product of virgins, is symbolic of purity.

The high cost of beeswax, as opposed to paraffin and other waxes, prohibits it being more widely used industrially; however, the beekeeper who has access to a solar wax extractor or, better still, a wax press may find it economical to use in several homemade products. The most common articles made at home are candles and furniture and car polishes. Some beekeepers prepare small, one- or two-ounce cakes of wax that they offer for sale along with their honey.

In recent years advertisements for foundation presses and mills have appeared in some of the bee journals; I question whether it is economical for beekeepers in North America to make their own foundation, but in many countries it is reasonable to do so. One of the chief problems with making one's own foundation is that it is invariably too thick. This does no harm insofar as the bees using the foundation are concerned, but it does take more wax; one may argue that this wax may be recovered later when the comb is rendered, but that may be only after many years.

Candles made at home are best made of capping wax, as they will have a bright yellow appearance. Both molded and dipped candles are popular and equipment for making both is advertised in the bee journals. The selection of the wick is especially important. When a candle is lit, liquid wax is carried up the wick by capillary action. Candle wicking is braided and one of the threads is pulled tighter than the others so that the tip will curl in the flame. If the wick stands up straight, the end will char, but not burn completely; the charred portion will prevent the wax from rising farther, causing the candle to smoke. Wicking varies in size and shape according to the type and diameter of the candle. It is important to select the proper wicking for the type of candle one wants to make.

Polishes may be made by combining beeswax and turpentine, the proportions varying according to whether one wants a thick or a thin paste. Soap is recommended in some formulas. The odor of turpentine is not too pleasant and adding a perfume may be helpful.

A word of caution is in order: One should never heat wax over an open flame! Beeswax contains some water and when the melted wax reaches the boiling point of water, it may foam up over the edge of the container and burst into flame. It is safest to heat wax with steam. It is also simple to liquify wax in a solar wax extractor.

OTHER HIVE PRODUCTS

Besides honey, several items have been harvested and marketed from beehives; some have well-documented special virtues, but many of the claims for these products are vague and unsubstantiated.

[Propolis is the name given to the tree gums and resins that honeybees collect to varnish the hive interior and fill cracks and crevices; it is an excellent waterproofing agent. Propolis also contains natural substances that will kill or prevent the growth of microbes; it gives the same protection against infection in the beehive that it gives a tree wound. People have used propolis in polishes and varnishes, and it was probably used by the ancient Egyptians for embalming. There are claims that propolis was used in the manufacture of Stradivarius violins, but this cannot be proven. In recent years there have been a variety of articles, mostly from Eastern Europe, recommending propolis and/or propolis extracts in dermatology and in the treatment of ulcers, diseases of the teeth and gums, and a wide variety of other medical problems.] Many chemical analyses have been made, with variable results, depending greatly on the kinds of plants from which the bees have collected. There are no indications that honeybees add anything to propolis. Bees have been known to collect road tar, wet paint, and caulking compound in areas where natural propolis is unavailable. If such substances are included in "propolis" harvested from a hive, any medical use would be hazardous.

With an electrical shocking device designed by Dr. Allen Benton when he was a graduate student at Cornell University, the venom from approximately 50,000 bees may be collected in two hours. As a result of Benton's work, pure venom is available to physicians for desensitizing that very small fraction of our population who are severely allergic to bee stings. In addition, some books and many articles have been written stating that bee venom, either from actual stings or injections of the collected product, is valuable in the treatment of arthritis and rheumatism. One recent medical review stated that there are too many favorable case histories and testimonials to dismiss these claims completely. This is an area where there is no good documentation and research on the subject is needed. In any case, honeybee venom is so easily collected that currently one beekeeper alone easily supplies the entire world demand, so there is little (if any) room for monetary gain for newcomers.

[Royal jelly is the name given to a glandular secretion produced by young worker bees and used as larval food. Those larvae destined to become queens are fed royal jelly exclusively, whereas those that will become worker bees receive royal jelly for two days of larval life, after which they are fed pollen and honey. It is also said that royal jelly is fed to adult queens, but while this may be so, and it certainly sounds reasonable, it has never been demonstrated.] In the late 1940s and 1950s there were many claims, mostly from Europe and especially Eastern Europe, that since royal jelly can make the difference between a worker and a queen bee, it might have a special virtue in human nutrition. It was also claimed to have "rejuvenating" powers. Such claims have never been proven; however, in many Asian and European countries one may buy royal jelly for various "health" and cosmetic purposes in the form of pills, capsules, salves, creams, and even as an injection.

Another fad is the use of pollen in human nutrition. Claims for its special value appear to come from many of the same people who advocate the use of other hive products in medicine. Again, the data are scant.

There is no doubt in my mind that many claims for the medical value of certain hive products have their basis in the fact that honeybees are truly amazing animals. They live successfully and, for the most part, free from harmful microbes in an intensely crowded environment where disease might spread easily. I think, too, that the highly organized state of a beehive, as well as its cleanliness and thriftiness, cause us to see in honeybees virtues we generally admire in ourselves. This leads one to think that what is good for a bee is good for a human, a parallel that must be considered with great caution.

Great progress in understanding bee biology has been made in recent years. Of special importance has been the development of tools in the field of chemistry that allow one to collect, analyze, and identify minute quantities of materials, even down to one part in a billion. Many mysteries of life on earth have been illuminated, and studies on honeybees have contributed to our increasing understanding of the processes of life. While much remains to be explained, man continues to chip away at unknowns every day.

SELECTED BIBLIOGRAPHY

JOURNALS

There are three monthly nationwide bee journals, the first two of which, below, have been in circulation for over one hundred years. These report current events, list upcoming meetings, offer beekeeping supplies, and have timely articles on bee management. Single copies are sent free on request. These are:

American Bee Journal, Dadant and Sons Inc., Hamilton, Illinois 62341

Gleanings in Bee Culture, A. I. Root Co., P.O. Box 706, Medina, Ohio 44258-0706

The Speedy Bee (newspaper format), P.O. Box 998, Jesup, Georgia 31545

BOOKS

Anatomy

Anatomy of the Honey Bee. R. E. Snodgrass. Ithaca, N.Y.: Cornell University, 1956. 330 pages.

Anatomy and Dissection of the Honeybee. H. A. Dade. London: Bee Research Association, 1962. 158 pages.

Behavior

The Social Behavior of the Bees. Charles D. Michener. Cambridge, Mass.: Belknap Press, 1974. 404 pages.

Bees: Their Vision, Chemical Senses and Language. Karl von Frisch. Ithaca, N.Y.: Cornell University Press, 1971, revised edition. 157 pages.

The Dance Language and Orientation of Bees. Karl von Frisch. Cambridge, Mass.: Belknap Press, 1967. 566 pages.

Cooking with Honey

The Honey Kitchen. Hamilton, Ill.: Dadant and Sons, 1981. 208 pages.
The Wonderful World of Honey, A Sugarless Cook Book. Joe Parkhill. Kansas City, Mo.: Cook Book Publishers, 1978. 160 pages.

Comb Honey

Comb Honey Production. Roger A. Morse. Ithaca, N.Y.: Wicwas Press, 1978. 128 pages.
Fifty Years Among the Bees. Dr. C. C. Miller. 1915. Reprinted Molly Yes Press (New Berlin, N.Y.), 1980.
Honey in the Comb. Eugene E. Killion. Hamilton, Ill.: Dadant and Sons, 1980. 148 pages.
How to Raise Beautiful Comb Honey. Richard Taylor. Interlaken, N.Y.: Author, 1977. 75 pages.

Diseases

Honey Bee Pests, Predators and Diseases. Roger A. Morse, ed. Ithaca, N.Y.: Cornell University Press, 1978. 419 pages. Detailed discussion of bee diseases, written by sixteen specialists.
Honey Bee Brood Diseases. Henrik Hansen. Ithaca, N.Y.: Wicwas Press, 1980. 32 pages. Includes color pictures.
Identification and Control of Honey Bee Diseases. H. Shimanuki, Superintendent of Documents. Washington, DC.: U.S. Government Printing Office, 1977. 18 pages.
Honey Bee Pathology. Leslie Bailey. New York: Academic Press, 1981. 124 pages.

General Texts

Langstroth on the Hive and the Honey-bee. L. L. Langstroth. Medina, Ohio: A. I. Root Co., 1978. 402 pages.

ABC and XYZ of Bee Culture. A. I. Root, E. R. Root, H. H. Root, and J. A. Root. Medina, Ohio: A. I. Root Co., 1975. 726 pages.
The Hive and Honey Bee. Dadant and Sons, eds. Hamilton, Ill: Dadant and Sons, 1975. 740 pages.
Bees and Beekeeping. Roger A. Morse. Ithaca, N.Y.: Cornell University Press, 1975. 295 pages.
The Joys of Beekeeping. Richard Taylor. New York: St. Martin's Press, 1974. 166 pages.

Honey

Honey. Eva Crane, ed. New York: Crane, Russak and Company, 1975. 608 pages.

Honey Plants

American Honey Plants. Frank C. Pellett. Hamilton, Ill.: Dadant and Sons, 1977. 497 pages.
Honey Plants Manual. Harvey B. Lovell. Medina, Ohio: A. I. Root Co., 1956. 64 pages.

Pollination

Insect Pollination of Cultivated Crop Plants. S. E. McGregor. Washington, D.C.: Agricultural Research Service, USDA, 1976. 411 pages.
Insect Pollination of Crops. John B. Free. New York: Academic Press, 1970. 544 pages.

Queen Rearing

Rearing Queen Honey Bees. Roger A. Morse. Ithaca, N.Y.: Wicwas Press, 1979. 128 pages.
Contemporary Queen Rearing. Harry H. Laidlaw, Jr. Hamilton, Ill.: Dadant and Sons, 1979. 212 pages.
Instrumental Insemination of Honey Bee Queens. Harry H. Laidlaw, Jr. Hamilton, Ill.: Dadant and Sons, 1977. 144 pages.

INDEX

agricultural colleges, 42–43
almond pollination, 79
aluminum reinforced foundation, 31
American Bee Journal, 161
American Beekeeping Federation, 44
American foulbrood, 2, 50, 53–54, 117–18, 145–47
American Honey Producers Association, 44
anatomy, 161
apiaries
  inspection, 43
  sites, 13, 15–16
Arnaba Ltd., 31–32
associations for beekeepers, 43–44
Avitabile, A., 16
Bailey, L., 163
bait hives, 1, 9–11
baking with honey, 152
banning bees, 13
bears, 148–51
bee(s)
  blowers, 109–10
  books, 44–45, 161–64
  catalogs, 45–46
  diseases, 12, 43, 53–54, 117–18, 120–21, 162
  equipment, 45–46
  escapes, 108–9
  journals, 44–45, 161
beekeepers' associations, 43–44
beekeepers' newsletters, 43–44
beekeeping areas, 15
beekeeping literature, 42–43
bee repellents, 107
bees in buildings, 7–9
bee space, 19, 24
bee stings, 38–39
beeswax, 137, 140–41
  candles, 156–57
  polish, 157
  products, 156–57
bee trees, 7–9, 126
bee venom, 158–59
behavior, 162
Benton, A., 158
benzaldehyde, 107
Bioenvironmental Bee Laboratory, 145
bird feeders and bees, 46
black bees, 12
blueberry pollination, 80
Boardman feeders, 56
books and journals, 44–45
bottomboards, 17–18, 20, 33–34
bottom supering, 91
brace comb, 32, 80, 107
brood, 2, 16, 58
brood nest, 52
brood nest temperature, 4, 16, 142
brood patterns, 58
brood rearing, 52
brood rearing temperature, 4, 16, 142
brushing bees, 110, 121
building paper pack, 49–51, 132
bulletins on bees, 42–43
burr comb, 32, 57, 80, 107
butyric anhydride, 107–8
buying established colonies, 2–3
buying packages, 3–5
calming bees, 83
Canada, beekeeping in, 4–5, 18, 41, 62, 129
candle wicking, 157
capturing swarms, 5–6, 9–11
carbon dioxide in colonies, 18
Carniolan bees, 11–12
catalogs, 45–46
Caucasian bees, 11–12, 14
chalkbrood, 146–47

INDEX

chilled brood, 148
circulars on bees, 42–43
city beekeeping, 13–15
cleansing flights, 50
clipping queens, 78–79
clothing, protective, 35–37
coal dust, 46
Cobana comb honey sections, 31
Coggshall, W. L., 31
cold knives, 112
collecting swarms, 5–6
colonies in snowbanks, 18
colonies
  growth, 68
  registration, 43
  ventilation, 50, 82
colors bees see, 35
comb honey production, 91, 102–4, 137, 162
Combs, G. F., 63
combs, making, 100
compensation programs, 85
congestion of colonies, 69–72
contaminated pollen, 85
control of bee diseases, 53–54
cooked honey, 154–56
cooking with honey, 162
cooling hives, 34
correspondence courses, 45
cotton clothing, effect on bees of, 35
Crane, E., 163
cranberry pollination, 80
creosote, 34
crocuses, 48
cut-comb honey production, 102–4, 137
cutting bee trees, 7–9
Dadant and Sons, Inc., 161–63
Dade, H. A., 162
DDT, 84
dead bees in combs, 50–51
dead bees on the snow, 18
Demaree, G. W., 77
Demareeing, 77
design of beekeeping equipment, 20–21
diagnosing bee diseases, 43
dimensions for bee equipment, 24
diseases, 43, 53–54
  control, 43, 162
  disease-resistant queens, 12
  inspection, 117–18, 120–21
division board feeders, 56
drawing new foundations, 89–91

dressing for the apiary, 35–37
drifting, 34–35
drugs for bee diseases, 53–54
dry swarms, 5–6
Dyce, E. J., 133
Dyce Laboratory, 47
Eastern Apicultural Society, 44, 138
eight-frame hives, 21
entrances
  cleats, 21
  feeders, 56
  reducers, 21
  plugged, 40
Environmental Protection Agency, 85–86
equalizing colonies, 58–59
equipment, quality of, 21
equipment, standardization of, 24
European foulbrood, 11–12, 145–47
excluders, 90, 100
extension entomologists, 42–43
extracting honey, 89–91, 110–12, 155
factory-made equipment, 21
fall feeding, 127–29, 135
feces, 14–15
feeding bees, 55–58, 127–29
fermentation, 139
Food and Drug Administration, 53–54
food reserves, 54
foundations, 3, 29, 64, 89, 91
  plastic, 32
  reinforcing, 31
frames, 24, 27–30
free-hanging frames, 30
Free, J. B., 164
Frisch, K. von, 35, 162
fuel for smokers, 38
fungus disease, 146
German black bees, 12
Gleanings in Bee Culture (journal), 43, 161
gloves, 37
graham crackers, honey in, 152
granulation, 139
growing queens, 12
guard bees, 37
Hansen, H., 163
harvesting, 107–10
  comb honey, 114
  cut-comb honey, 114
  fall honey, 121
hives, 19–27

INDEX • 165

homemade extractors, 112
honey-bee life history, 142
honey
    extracted, 155
    fermentation, 139
    filtered, 155
    flows, 101
    jelly, 152–53
    organic, 154–56
    plants, 47, 163
    products, 152
    shows, 136
    vinegar, 153–54
    wine, 153–54
hot rooms, 112–13
hunting bee trees, 126
ice for bee stings, 39
ice on bottomboards, 17
inspecting colonies, 69–72
installing package bees, 59–67
Italian bees, 11–13
Jeffree, E., 16
judging honey, 136–38
judge's scorecard, 138
keeping bees in cities, 13–15
khaki clothing, effect on bees of, 35
Killion, E., 162
labeling honey, 139–40
labels on pesticides, 86
Laidlaw, H. H., Jr., 164
Langstroth, L. L., 19, 163
Langstroth dimensions, 21
Langstroth super, 19–20
lead arsenate, 85
leather, effect on bees of, 35
life history of the honey bee, 142
light, effect on bees of, 10–11
literature on bees, 42–43
London purple pesticide, 84
Lovell, H. B., 163
making increase, 93–95
marketing honey, 104
marking queens, 78–79
mating in honey bees, 12, 143–44
McGregor, S. E., 163
mead, 153–54
metal comb, 31
metal eyelets, 29
mice, 122, 135, 148–51
Michener, C. D., 162
Miller, C. C., 104, 162
Miller-type feeders, 56
modern beehives, 19
moisture in honey, 105–6, 113–14

mold in bee food, 55
moldy combs, 50–51
mothballs, 96
mother of vinegar, 153
moving bees, 80
nailing beekeeping equipment, 21
nectar, 68
    nectary, 101
    production, 89
    secretion, 101
newsletters, 43–44
nosema disease, 54
nucleus colonies, 94, 115
number of colonies in an apiary,
    15–16
odors, effect on bees of, 35
old comb, effects of, 92–93
package bees, 1, 3–5, 59–67
packing honey, 138–39
painting equipment, 32
Paris green pesticide, 84
paradichlorobenzene, 96
Parkhill, J., 162
pasteurizing honey, 139
Peace River district, Can., 4, 129
Peer, D., 129
Pellett, F. C., 163
Penncap-M, 85
Pennwalt Corporation, 85
pentachlorophenol, 32–34
pesticides, 84–87, 147
plans for bee equipment, 24
plastic beekeeping equipment,
    30–32
plastic foundation, 32
polishes with beeswax, 137
pollen
    collection of, 41, 46–48
    contamination, 85
    plants, 47
    substitutes, 48
    supplements, 48
pollination, 79–80, 163–64
preparation for winter, 129–34
propolis, 12, 19, 57, 158
protective clothing, 35–37
Pure Food and Drug Administration,
    53–54
pure honey, 155
queen cage candy, 64
queen cups, 73, 76
queen excluders, 20, 97–99, 118
queen failure, 58
queen finding, 94

queen rearing, 164
races of bees, 11–13
raw honey, 154–56
refractometers, 106
registration of colonies, 43
removing bees
  from buildings, 7–9
  from supers, 107–10
  from trees, 7–9
rendering cappings, 140–41
rendering old combs, 140–41
renting bees for pollination, 79
repellents, 114
repellent pads, 107
replacing queens, 2
requeening, 115–17
Root, A. I., 163
Root Co., A. I., 161, 163
Root, E. R., 163
Root, H. H., 163
Root, J. A., 163
round-comb honey sections, 30–31, 137
royal jelly, 159
sacbrood, 54, 146
sawdust, 46
scale hives, 67
score card for honey, 138
screening bees, 82
secondhand colonies, 1–3
Seeley, T. D., 9–10
selecting the apiary site, 13
Sevin, 85
shaking bees, 110, 121
Shimanuki, H., 163
short courses, 45
skunks, 148–51
small cells, effect of, 92
smoke, effect of on bees, 37–38
smokers and fuel, 37–38
smoking bees, 69
Snodgrass, R. E., 161
snow, effect of on bees, 18
solar wax extractors, 43, 141
sorting combs, 100
Southern States Beekeeper's Organization, 44
Speedy Bee, 161
stainless-steel extractors, 111
standardization of equipment, 24
state colleges, 42–43, 45

steam knives, 111–12
stings, protection from, 38–39
stress diseases, 147
suede, effect of on bees, 35
sugar syrup, 14, 56–58
supering, 91–92
supers, 20
supersedure, 12
swarm boards, 117
swarming, 48, 72–80, 97–98
swarms, 1, 5–6
  collecting, 5–6
  prevention, 72–76, 117
sweat, effect of on bees, 35–36
Taylor, R., 162–63
top supering, 91
Townsend, G. F., 93
toxicity of pesticides, 83–87
two-queen system, 49
uncapping combs, 110–12
uncapping knives, 111–12
Union Carbide Corporation, 85
uniting weak colonies, 51–53, 122
unpacking colonies, 49–51
unripe honey, 68
upper entrances, 16, 99–100
used equipment, 2
veils, 35
venom, bee, 38–39
venom collection, 158–59
ventilation, colony, 82
wasps, 124–25
water, quality of for bees, 14
water to cool the hive, 14
wax extractors, 141
wax moths, 95–96, 103
wax presses, 141
weak colonies, 122
weighing colonies, 127
Western Apicultural Society, 44
wing clipping, queen, 78–79
winter flights, 16
wintering bees, 16–18, 49–51, 91, 127–34
wiring frames, 29–30
wood preservatives, 32–34
wool, effect of on bees, 35
wrapping bees for winter, 132–34
yellow jackets, 124–25
zinc queen excluders, 98
Zbikowski, W., 31